PREFACE

Chapter 1 - The Corps is responsible for dredging sediment from waterways to maintain shipping routes important for commerce. One dredge type, a hopper dredge, performs much of the dredging in ports and harbors, and the Corps uses its own fleet of hopper dredges and contracts with industry to carry out the work. In 2003, GAO examined the Corps' hopper dredging program and made recommendations to improve its management. GAO was asked to review changes to the program.

This report examines (1) actions the Corps has taken to address GAO's 2003 recommendations for improving the information needed to manage its hopper dredging program and develop cost estimates for industry contracts; (2) effects since 2003, if any, of the statutory restrictions placed on the use of the Corps' hopper dredges; and (3) key challenges, if any, the Corps faces in managing its hopper dredge fleet. GAO reviewed laws, regulations, and policies governing the Corps' use of hopper dredges, and related Corps reports. GAO analyzed dredging contract and financial data for fiscal years 2003-2012, assessed the reliability of these data, and interviewed Corps and dredging stakeholders.

Chapter 2 - The fiscal year 2002 Conference Report for the Energy and Water Development Appropriations Act directed GAO to study the benefits and effects of the U.S. Army Corps of Engineers' (Corps) dredge fleet. GAO examined the characteristics and changing roles of the Corps and industry in hopper dredging; the effect of current restrictions on the Corps' hopper dredge fleet; and whether existing and proposed restrictions on the fleet, including the proposal to place the *McFarland* in ready reserve, are justified. In addition, GAO identified concerns related to the government cost estimates the Corps prepares to determine the reasonableness of industry bids.

Chapter 3 - The Conference Report for the 2004 Energy and Water Appropriations Act directed the Corps of Engineers to report to the Appropriations Committees with a detailed plan of how it intends to rectify the issues raised in the March 2003 GAO report entitled "Effects of Restrictions on Corps' Hopper Dredges Should be Comprehensively Analyzed". The plan is to include how the Corps intends to establish a baseline for determining the appropriate use of the Corps hopper dredge fleet in the future. The conferees also requested that the Corps include a comprehensive analysis of the costs and benefits of the existing and proposed restrictions on the use of the fleet, expected the Corps to put in place measures by which better investment decisions regarding the fleet can be made.

The goal of this analysis was to identify the best business case that can ensure the ability to provide an acceptable level of service to the ports and users of the Nation's waterborne transportation system. The degree of change and configuration of the Corps minimum hopper dredge fleet was determined by resolving an acceptable level of risk both to the industry and to the ports, Corps partners and maritime users. In order to ensure that the industry dredges are consistently available to support the dredging requirements of the Navigation Program, there must be a reasonable ability to realize a return on investment that will keep the vessels and equipment fully operational.

This report explains the navigation mission, the dredging environment, current problems, and issues concerning management of the four hopper dredges in the Corps minimum dredge fleet and the combined industry and Corps hopper dredge fleet. The data management issues raised in the GAO report are addressed and the improvements described. The analysis of costs and benefits of existing and proposed restrictions on the use of Corps hopper dredges is presented and the results are presented in terms of evaluations of twelve options.

The recommended option would schedule the Corps hopper dredge *Essayons* for 215 days, schedule the *Yaquina* for 178 days, keep the *Wheeler* in ready reserve, retire the *McFarland,* and continue with the initiatives from the Industry Corps Hopper Dredge Management Group. This Option would result in a $10.1 million net reduction in the total cost for hopper dredging, offer approximately 55 days additional work to industry, ensure that there is a viable reserve capability ready to respond to unforeseen requirements, and ensure that Federal projects requiring hopper dredging can be accomplished in a timely manner and at reasonable cost.

In: Hopper Dredging of Waterways
Editor: Eleanor G. Chance
ISBN: 978-1-61122-357-6

Chapter 1

ARMY CORPS OF ENGINEERS: ACTIONS NEEDED TO FURTHER IMPROVE MANAGEMENT OF HOPPER DREDGING*

United States Government Accountability Office

WHY GAO DID THIS STUDY

The Corps is responsible for dredging sediment from waterways to maintain shipping routes important for commerce. One dredge type, a hopper dredge, performs much of the dredging in ports and harbors, and the Corps uses its own fleet of hopper dredges and contracts with industry to carry out the work. In 2003, GAO examined the Corps' hopper dredging program and made recommendations to improve its management. GAO was asked to review changes to the program.

This report examines (1) actions the Corps has taken to address GAO's 2003 recommendations for improving the information needed to manage its hopper dredging program and develop cost estimates for industry contracts; (2) effects since 2003, if any, of the statutory restrictions placed on the use of the Corps' hopper dredges; and (3) key challenges, if any, the Corps faces in managing its hopper dredge fleet. GAO reviewed laws, regulations, and policies governing the Corps' use of hopper dredges, and related Corps

* This is an edited, reformatted and augmented version of the United States Government Accountability Office publication, GAO-14-290, dated April 2014.

reports. GAO analyzed dredging contract and financial data for fiscal years 2003-2012, assessed the reliability of these data, and interviewed Corps and dredging stakeholders.

What GAO Recommends

GAO recommends the Corps provide written direction to its district offices on consistently populating its database with no-bid and high-bid solicitations and develop a written plan for a study to obtain and periodically update certain hopper dredging cost data for its cost estimates. The Department of Defense concurred with the recommendations.

What GAO Found

The U.S. Army Corps of Engineers (Corps) has taken actions to address GAO's 2003 recommendations for improving information related to hopper dredging, but some data gaps remain. First, in response to GAO's recommendation to obtain and analyze data needed to determine the appropriate use of its hopper dredge fleet, the Corps established a tracking log to document urgent or emergency work its dredges carry out. The Corps also modified its dredging database to track solicitations for industry contracts that received no bids and bids exceeding the Corps' cost estimate by more than 25 percent, referred to as high bids. Corps district offices, however, do not consistently enter data on these solicitations, and Corps headquarters has not provided written direction to the district offices to ensure data are consistently entered. Tracking and analyzing no-bid and high-bid solicitation data could enable the Corps to identify and address gaps in industry's ability to fulfill certain dredging needs as the Corps plans its future hopper dredging work. Second, in response to GAO's recommendation, the Corps took action to assess the data and procedures it used for developing cost estimates when soliciting industry contracts. However, certain industry cost data the Corps relies on remain outdated. For example, some of the data it uses on hopper dredge equipment date back to the late 1980s. A senior Corps official stated that a study could be conducted to update the data, but the Corps has no plans to conduct such a study. Having a plan for obtaining updated data is important for developing sound cost estimates.

Statutory restrictions on the use of the Corps' hopper dredges since 2003 have resulted in costs to the Corps, but the effect on competition in the hopper dredging industry is unclear. Restrictions limiting the number of days that Corps dredges can work have resulted in additional costs such as costs to maintain certain Corps dredges while they are idle; the Corps incurs many of the costs for owning and operating its hopper dredges regardless of how much they are used. The restrictions, however, help ensure the Corps has the ability to use these dredges to respond to urgent or emergency dredging needs when industry dredges are unavailable. It is not clear to what extent restrictions have affected competition in the dredging industry. The number of U.S. companies with hopper dredges has not changed, but the number and size of these dredges have decreased since 2003. In addition, GAO did not find evidence of increased competition based on the number of bidders and winning bid prices for Corps hopper dredging projects since 2003.

Key challenges facing the Corps in managing its hopper dredge fleet are (1) ensuring the fiscal sustainability of its hopper dredges and (2) determining the fleet's appropriate future composition. In 2012, the Corps determined that because of increasing ownership and operating costs, among other things, its hopper dredges would become unaffordable unless actions were taken, including increasing the daily rates charged to projects using the Corps' dredges. Factors such as the aging of the Corps' fleet and the effect on industry of possible changes to the Corps' fleet make it difficult for the Corps to determine the best fleet composition. In studies it conducted in 2011 and 2012, the Corps identified actions that could help address these challenges, such as reviewing the operating costs of hopper dredges to evaluate the affordability of certain dredges.

ABBREVIATIONS

DCA Dredging Contractors of America

April 10, 2014

The Honorable Timothy H. Bishop
Ranking Member
Subcommittee on Water Resources and Environment
Committee on Transportation and Infrastructure
House of Representatives

Dear Mr. Bishop:

Maintaining the nation's navigation channels and ports is vital to U.S. commerce, with nearly $1.8 trillion of import and export cargo passing through the nation's waterways in 2012. The U.S. Army Corps of Engineers (Corps) is responsible for dredging—that is, removing sediment from the bottom of—ports, harbors, and other U.S. waterways to facilitate their navigation. Vessels called dredges are used to remove sediment in order to maintain navigation channels at the depths and widths necessary for shipping. Various types of dredging vessels exist, each designed to perform in certain conditions or locations. One type—the hopper dredge—performs much of the dredging work in ports, harbors, and other waterways exposed to the ocean, where shipping traffic and operating conditions render the use of other types of dredges inefficient, impractical, or unsafe. A hopper dredge vacuums material (e.g., a mixture of sediment and water) into its "hopper" or containment area, where the material is stored before being transported to a placement site (see figure 1). In fiscal year 2012, the Corps spent about $370 million for hopper dredging, removing nearly 72 million cubic yards of material from dozens of ports and harbors across the country.

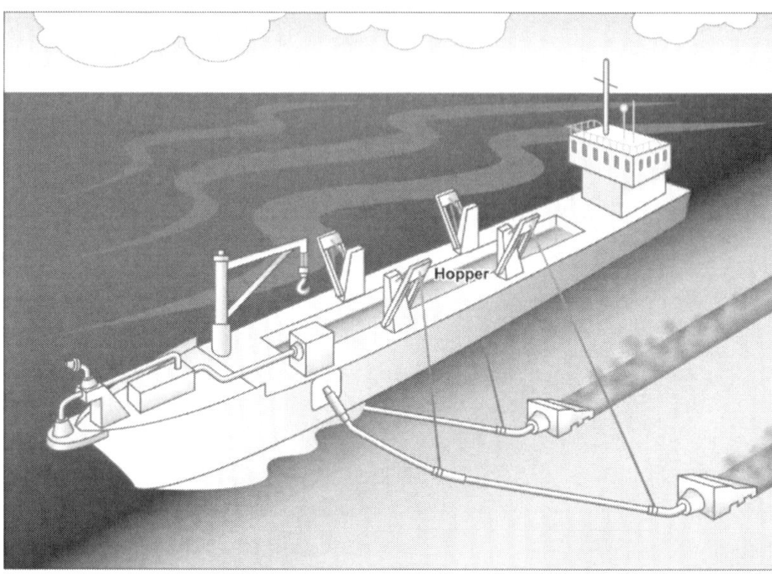

Source: GAO.

Figure 1. Illustration of a Hopper Dredge.

The Corps carries out hopper dredging using a combination of its own fleet of dredging vessels and industry vessels operating under contract, and it is required to do so in the manner most economical and advantageous to the United States. Until 1978, the Corps performed all hopper dredging in the United States with its fleet of 14 hopper dredges. In 1978, legislation directed the Corps to (1) contract out much of its dredging work to private industry as industry demonstrated that it could perform the work at reasonable prices and in a timely manner, (2) reduce the federal hopper fleet as industry demonstrated its ability to perform, and (3) maintain a minimum fleet of federal vessels.[1] Specifically, the legislation directed the Corps to retain as much of its fleet as it determined necessary to ensure that the federal and private industry fleets together could carry out necessary dredging projects. According to the Corps, the purpose of the minimum fleet is to diminish risks to navigation by performing urgent, emergency, or national defense work, or to respond when industry either does not submit a reasonable bid for dredging solicitations or does not perform adequately under an existing contract.[2] Over time, the Corps has reduced the size of its hopper fleet to four vessels: the *Essayons*, *McFarland*, *Wheeler*, and *Yaquina*. Generally, Corps district offices, in coordination with Corps division offices to which the district offices report, are responsible for managing and contracting for the dredging that occurs within their districts, under the guidance of Corps headquarters.

Beginning in 1993, Congress began imposing various statutory restrictions on the operations of certain Corps hopper dredges. For example, restrictions imposed by the Water Resources Development Act of 1996 effectively (1) reduced the annual authorized work schedule of the *Essayons*, *McFarland*, and *Yaquina* and (2) limited the number of days the *Wheeler* could be used annually by taking it out of active status and placing it in "ready reserve"—under which the Corps uses it primarily for training purposes and calls it into active status only in limited circumstances, such as urgent and emergency work.[3] According to a House committee report, the restrictions were intended to further encourage the Corps to contract with private industry for hopper dredging.[4] Over a decade later, Congress passed the Water Resources Development Act of 2007, which required that the Corps place the *McFarland* in ready reserve and limited the use of the vessel's work.[5] In contrast, the 2007 act lifted the restrictions that had been in place on the number of days the *Essayons* and *Yaquina* could work

In a review we conducted in 2003, we found that the restrictions in place at that time had imposed costs on the Corps' dredging program but had not resulted in proven benefits, such as increased industry competition or lower prices for hopper dredging.[6] In our 2003 report, we made three recommendations to the Corps for improving the information used to manage its hopper dredges and develop cost estimates for industry contracts. Specifically, we recommended that the Corps

- obtain and analyze baseline data needed to determine the appropriate use of its hopper dredge fleet, such as data on emergency work performed by the Corps and solicitations that receive no bids;
- assess the data and procedures used to perform the cost estimate used when contracting dredging work to the hopper dredging industry; and
- prepare a comprehensive analysis of the costs and benefits of existing and proposed restrictions on the use of its hopper dredge fleet.

The Corps concurred with our recommendations and agreed to implement them.

You asked us to examine changes to the Corps' hopper dredging program since our review in 2003. This report examines (1) the actions the Corps has taken to address our 2003 recommendations for improving the information needed to manage its hopper dredging program and develop cost estimates for industry contracts; (2) the effects since 2003, if any, of the statutory restrictions placed on the use of the Corps' hopper dredges; and (3) key challenges, if any, the Corps faces in managing its hopper dredge fleet.

To examine the actions the Corps has taken to address our 2003 recommendations, we reviewed the Corps' process for tracking and analyzing data on solicitations for private industry hopper dredging work, several sources of information the Corps maintains on the use of its hopper dredge fleet, the Corps' 2005 report to Congress on hopper dredges, and Corps documentation related to developing government cost estimates. To examine the effects since 2003, if any, of the statutory restrictions placed on the use of the Corps' hopper dredges, we analyzed the statutes and regulations governing the use of the Corps' hopper dredge fleet, reviewed Corps reports and financial data on the fleet, obtained and reviewed information from the dredging industry on its fleet of hopper dredges, and analyzed dredging data collected by the Corps through its Dredging Information System. To assess the reliability of the data, we interviewed Corps officials who maintain the database, reviewed related documentation, and tested the data for missing or erroneous values. We determined that the data we used were sufficiently reliable for our purposes. To examine key challenges, if any, the Corps faces in managing its hopper dredge fleet, we reviewed Corps studies and financial data on its hopper dredge fleet. For all three objectives, we interviewed officials from Corps headquarters, division offices, and the 9 Corps district offices with the largest hopper dredging workload during fiscal years 2003 through 2012 (out of a total of 17 district offices that contracted with industry for hopper dredging work during the time period). We also interviewed representatives from the five dredging companies that own and operate hopper dredges, and other stakeholders including local pilots' associations and port authorities from the areas where the Corps' hopper dredges are stationed, along with their national-level counterparts. In addition, we visited the Corps' four hopper dredges and one industry hopper dredge for informational tours of these vessels to gain a better understanding of their physical characteristics and operations. Appendix I describes our objectives, scope, and methodology in greater detail.

We conducted this performance audit from January 2013 to April 2014 in accordance with generally accepted government auditing standards. Those standards require that we plan and perform the audit to obtain sufficient, appropriate evidence to provide a reasonable basis for our findings and conclusions based on our audit objectives. We believe that the evidence obtained provides a reasonable basis for our findings and conclusions based on our audit objectives.

BACKGROUND

Since 1824, the Corps has been responsible for maintaining a safe, reliable, and economically efficient navigation system in the United States. This system currently comprises more than 12,000 miles of inland and intracoastal waterways and about 180 ports handling at least 250,000 tons of cargo per year. The accumulation of sediment in waterways—known as shoaling—reduces their navigable depth and, without dredging, may result in restrictions on vessels passing through the waterways. These restrictions often apply to the vessels' draft—the distance between the surface of the water and the bottom of the hull—which determine, in part, the minimum depth of water in which a vessel can safely navigate. Draft restrictions may result in delays and added costs as ships may need to off-load some of their cargo to reduce their draft, wait until high tide or until waterways are dredged, or sail into another port. For example, according to a 2011 Corps study, 1 foot of shoaling in the lower Mississippi River could result in $2.8 billion worth of cargo being disrupted annually.[7] To minimize such risks to navigation, the Corps removed an annual average of about 229 million cubic yards of material from U.S. waterways from fiscal year 2003 through fiscal year 2012, at an average annual cost of about $1.1 billion, according to the Corps. Even with these efforts, draft restrictions have regularly been in place on major waterways throughout the United States in the past several years, according to Corps documents and officials.

The Corps contracts with industry to perform most dredging, including work done by hopper dredges. According to the Corps, of the approximately $11 billion it spent for dredging from fiscal year 2003 through fiscal year 2012, about $2.37 billion was for hopper dredging. Of that, industry hopper dredges accounted for about $1.8 billion, and Corps hopper dredges accounted for about $570 million. Corps spending on hopper dredging has more than doubled since fiscal year 2003, while the amount of material removed by hopper dredges has increased only slightly over that period, according to Corps data. Specifically, as shown in figureure 2, the Corps spent nearly $170 million for Corps and industry hopper dredges to remove around 66 million cubic yards of material in fiscal year 2003. By fiscal year 2012, Corps spending on Corps and industry hopper dredging had increased to about $370 million, while the amount of material removed increased to nearly 72 million cubic yards. This growth in spending reflects costs for hopper dredging that, according to Corps documents, have increased because of rising costs for fuel and steel, among other factors.

Hopper dredging today is generally performed in three regions of the United States—the East Coast, Gulf Coast, and West Coast—and each region has at least one Corps hopper dredge that typically operates in it: the *McFarland* on the East Coast, the *Wheeler* on the Gulf Coast, and the *Essayons* and *Yaquina* on the West Coast. On the East and Gulf Coasts, the majority of the hopper dredging workload is carried out by industry dredges, while on the West Coast, Corps dredges remove more than half of the dredged material. Various factors can influence and complicate hopper dredging in each region. For example, on the East Coast, much of the hopper dredging must be performed during certain months of the year because environmental restrictions related to endangered sea turtles and other species prohibit dredging while those species are present. On the West Coast, the Corps must factor in the time and expense of moving industry dredges through the Panama Canal if the only available industry hopper dredges are on the East or Gulf Coasts.

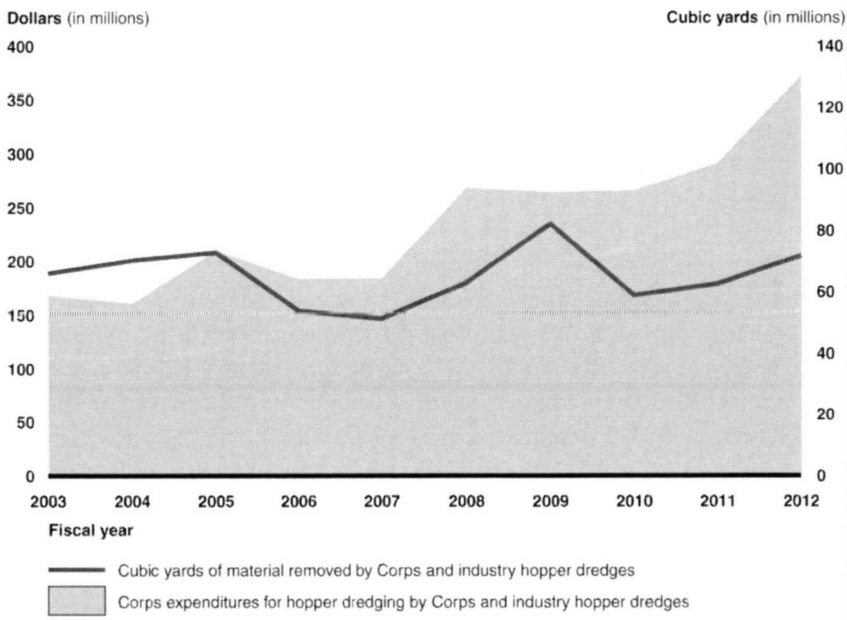

Source: U.S. Army Corps of Engineers.

Note: The Corps' hopper dredging work is funded through the Corps' annual civil works appropriation and, in most years since 2003, through supplemental appropriations, such as those for Hurricane Katrina recovery efforts, the American Recovery and Reinvestment Act, and other sources.

Figure 2. Corps Expenditures for Hopper Dredging and Amount of Material Removed by Corps and Industry Hopper Dredges, Fiscal Years 2003 through 2012.

The sizes and capabilities of specific hopper dredges—and, therefore, the projects for which they are suited—vary. For instance, shallow ports and harbors cannot be dredged by vessels with deep drafts in many cases. The Corps uses the *Yaquina*, which is a small dredge with a draft of around 15 feet when its hopper is fully loaded,[8] for dredging small and shallow ports along the California, Oregon, and Washington coasts. In contrast, the Corps uses the *Wheeler*, which is a large dredge with a draft of nearly 30 feet when its hopper is fully loaded, for deeper navigation channels such as those in the lower Mississippi River. See appendix II for a list of Corps and industry hopper dredges and their characteristics.

As noted, several pieces of legislation were enacted that sought to increase the role of industry in hopper dredging by placing restrictions on the use of the Corps' hopper dredges. More specifically, in 1978, legislation directed the

Corps to contract out much of its hopper dredging work to industry and reduce the Corps' fleet to the minimum necessary to insure the capability of the federal government and industry together to carry out projects for the improvement of rivers and harbors. The Energy and Water Development Appropriations Act for fiscal year 1993, and subsequent appropriations acts in the early 1990s, required the Corps to offer for competitive bidding at least 7.5 million cubic yards of hopper dredging work previously performed by the federal fleet. The Corps addressed this requirement by reducing the use of each of its four dredges from about 230 workdays per year to about 180 workdays per year. The Water Resources Development Act of 1996 then required the Corps to take the *Wheeler* out of active status and place it into ready reserve. The Corps implemented this requirement beginning in fiscal year 1998 by generally limiting the *Wheeler* to working 55 days a year plus any urgent or emergency work. More recently, the Water Resources Development Act of 2007 required that the Corps place the *McFarland* in ready reserve and limited the use of the vessel to 70 working days per year in the Delaware River and Bay, plus any urgent and emergency work.[9] See table 1 for the statutory restrictions in place on the use of the Corps' hopper dredges and how they have changed since fiscal year 2003.[10]

The Corps follows a process—known as the raise the flag procedure—for activating its ready reserve dredges to respond to urgent or emergency dredging needs. The Corps defines an urgent need for dredging as a time-sensitive situation that may require prompt action for providing a safe navigation channel, and an emergency as a situation that would result in an unacceptable hazard to life, a significant loss of property, or an immediate, unforeseen, and significant economic hardship if corrective action is not undertaken within a time period less than the normal contract procurement process. The raise the flag procedure includes a series of steps intended to allow industry the opportunity to respond to urgent or emergency dredging needs before the Corps uses its own dredges. The Corps district office with an urgent or emergency dredging need notifies the Corps division office overseeing it of the dredging need, and district and division staff review ongoing hopper dredging work under existing Corps contracts to see if any industry hopper dredges could be made available. If no industry hopper dredges could be made available, the offices notify Corps headquarters. The Corps' Director of Civil Works may then decide whether to use one of the Corps' ready reserve hopper dredges or make additional efforts to procure an industry dredge, such as by releasing a dredge from an existing contract.

Table 1. Statutory Restrictions and Other Information on the Use of the Corps' Hopper Dredges

	McFarland	Wheeler	Essayons	Yaquina
In ready reserve	Yes	Yes	No	No
Current limit on workdays	70 (in law)	70	None[a]	None[a]
Activities allowed within the limit	Routine dredging in the Delaware River and Bay to ensure its ability to perform urgent and emergency work	Routine dredging to ensure its ability to perform urgent and emergency work	Not applicable	Not applicable
Activities allowed beyond the limit	The Corps can call the dredge out of ready reserve for urgent or emergency dredging when industry dredges are unavailable	The Corps can call the dredge out of ready reserve for urgent or emergency dredging when industry dredges are unavailable	Not applicable	Not applicable
Changes since fiscal year 2003	Placed in ready reserve effective December 30, 2009	Change in workday limit from 55 to 70 effective fiscal year 2008	Removal of workday limit effective fiscal year 2009	Removal of workday limit effective fiscal year 2009

Sources: GAO analysis of selected legislation and Corps information.

[a] Restrictions imposed by the Water Resources Development Act of 1996 effectively limited the *Essayons* and *Yaquina* to about 180 workdays. The Water Resources Development Act of 2007 lifted these restrictions and these two vessels resumed operations without restrictions on October 1, 2008.

The Corps contracts for most of the hopper dredging work by soliciting competitive bids from industry. To determine the reasonableness of contractor bids, the Corps develops a government cost estimate for its hopper dredging solicitations. Government cost estimates are developed using information on the costs of owning and operating hopper dredges—including acquisition, fuel, and shipyard costs—along with information on the project for which the dredging is needed—including the amount and type of material to be removed, and the distance from the dredging site to the placement site. In soliciting bids from contractors, the Corps most commonly uses a sealed-bid process, through which it generally awards the contract to the lowest bidder with a bid that is no more than 25 percent above the government cost estimate.[11] If the Corps does not receive any bids or if all bids exceed the government cost estimate by more than 25 percent, the Corps may pursue a number of options, including (1) negotiating with bidders to get the bid within an awardable range of the cost estimate; (2) reviewing the cost estimate and revising it based on additional information, as appropriate, or (3) performing the work itself such as through its raise the flag procedure.

The costs to own and operate hopper dredges include costs such as payroll for the crews, fuel, repairs, and depreciation. Hopper dredging requires large capital outlays—a modern hopper dredge comparable in size to the *Wheeler*, for instance, would cost around $100 million to build, according to Corps and industry estimates—and related costs such as depreciation and replacement of engines or other major equipment can represent a relatively large portion of the dredges' total costs. The Corps and industry incur much of the costs for their hopper dredges—such as paying a crew and keeping engines and other systems in ready working condition—regardless of how much the dredges are used.

The Corps uses two funding sources from its annual civil works appropriation to pay for its hopper dredges. First, for the ready reserve vessels *McFarland* and *Wheeler*, funds are provided for each dredge to cover their costs while they are idle in ready reserve. Second, the Corps pays for the use of its dredges with project funds based on a daily rate it establishes for its dredges. According to Corps officials, the Corps sets a daily rate specific to each of its hopper dredges at least annually, based on factors such as the costs of owning and operating the dredge, and the amount of work the dredge is expected to perform. As the Corps uses its hopper dredges for projects, the Corps uses funds allocated for those specific projects to pay its dredges, based on the number of days its dredges work and the dredges' daily rate.

THE CORPS HAS TAKEN ACTIONS TO ADDRESS OUR 2003 RECOMMENDATIONS BUT DOES NOT COLLECT ALL OF THE RECOMMENDED DATA

Since our 2003 review,[12] the Corps has taken actions to address our recommendations for improving the information needed to manage its hopper dredging program and develop cost estimates for industry contracts, but some data gaps remain. First, the Corps collects data on urgent or emergency hopper dredging work but does not consistently collect the data that we recommended on solicitations that received no bids or where all the bids received exceeded the Corps' cost estimate by more than 25 percent. Second, the Corps assessed the data and procedures for performing the cost estimates it uses when soliciting industry contracts in response to our recommendation, but the Corps has not obtained updated data for some costs used in these estimates. Third, in response to our recommendation to prepare an analysis of the costs and benefits of existing and proposed restrictions on the use of its hopper dredge fleet, the Corps developed a report analyzing options on how to best maintain and operate its fleet.

The Corps Collects Recommended Data on Urgent or Emergency Dredging but Does Not Consistently Collect Data on Certain Solicitations

In response to our 2003 recommendation to obtain and analyze baseline data needed to determine the appropriate use of its hopper dredge fleet, the Corps established a tracking log as part of its raise the flag procedure to maintain and review urgent or emergency work its hopper dredges carry out, but it does not consistently collect certain solicitation information that we recommended. Having a means to track urgent or emergency dredging work helps the Corps ensure it is documenting and evaluating when and under what circumstances it will use its ready reserve dredges. According to Corps officials, the Corps established a tracking log in 2007 to systematically track information on the circumstances when urgent or emergency hopper dredging may be needed, and specifically when Corps' dredges would be used to meet those needs. Corps district offices that are faced with critical hopper dredging needs submit information on their plans to address the needs to their division and Corps headquarters for review and approval. The Corps' decision-making process for determining whether to use its ready reserve vessels is also documented via its tracking log. For example, in January 2013, a hopper dredge was needed to perform work along the North Carolina coast because certain areas had become severely shoaled and were impeding safe navigation. One industry bid was received to perform the work, but it exceeded the government cost estimate by more than 25 percent. After determining its cost estimate was reasonable, the Corps negotiated with the industry bidder in an attempt to get the bid within an awardable range of the Corps' cost estimate, but the parties were unable to come to an agreement. As a result, the Corps initiated its raise the flag procedure because of the urgent nature of the situation. Because no other industry contractors were available immediately to respond, the Corps used the *McFarland* to perform the dredging and documented its decision-making process in its tracking log.

We also recommended that the Corps obtain and analyze other data that could be useful in determining the appropriate use of the Corps' hopper dredges, including data on solicitations that receive no bids or where all the bids received exceeded the Corps' cost estimate by more than 25 percent. Corps officials we spoke with said that they are aware when a no-bid or high-bid situation occur, particularly when they use a Corps dredge through their raise the flag procedure because of such a situation. But by tracking and analyzing no-bid and high-bid solicitation data, the Corps may be better positioned to identify gaps in industry's ability to fulfill certain dredging needs—such as during certain times of the year, in particular geographic areas, or for particular types of projects—and avoid or address any gaps identified. In 2004, the Corps took steps to address our recommendation by modifying data fields in its dredging database, the Corps' database for maintaining dredging information on each of its dredging projects, to collect data on no-bid and high-bid solicitations. We found, however, that data for these solicitations were not consistently entered into the database across the Corps district offices responsible for entering it. In our review of the Corps' dredging database, we found that one district office entered data on no-bid and high-bid solicitations. Corps officials from several district offices told us that entering information into the database is tedious and time-consuming. They also indicated that they do not enter information for all data fields because the officials primarily use information from the database for planning and scheduling future dredging work, not for reviewing data on past solicitations or solicitations that did not result in an awarded contract, which would include no-bid and high-bid solicitations.

Corps headquarters officials we spoke with recognized that tracking and analyzing data on no-bid and high-bid solicitations is important and could serve as a useful decision-making tool in planning future hopper dredging work. However, they have not provided written direction to the district offices to help ensure data on these solicitations are consistently entered into the database. According to officials we spoke with, they have not done so because of other higher-priority action items. The officials added that they have made efforts to ensure district offices consistently enter accurate and complete data into the dredging database, such as emphasizing this activity during periodic meetings with district offices. These outreach efforts have been targeted at entering data into the dredging database as a whole, however, and have not focused specifically on the importance of the data field for tracking no-bid or high-bid solicitations, according to the officials. Federal internal control standards state that management should develop written policies and procedures that staff are to follow as intended.[13] Without complete data on no-bid and high-bid solicitations, the Corps may be missing opportunities to plan future hopper dredging work that identifies and addresses potential gaps in industry's ability to fulfill certain dredging needs based on this solicitation information.

The Corps Reviewed Data and Procedures for Performing Cost Estimates, but Some Information Remains Outdated

In response to our recommendation to assess the data and procedures used to perform the cost estimate used when contracting dredging work to the hopper dredging industry, the Corps took several actions to improve its cost estimates, but some of the information it relies on remains outdated, such as its dredge equipment cost information dating back to the late 1980s.[14] In 2004, and again in 2008, the Corps took actions to evaluate and update certain cost data used in its cost estimates. In 2004, the Corps prepared an internal document that summarized the steps it took to analyze, evaluate, and update certain cost data used in its cost estimates. For example, according to the document, the Corps examined repair and maintenance costs for industry hopper dredges and updated some data for dredge engines. In 2008, the Corps partnered with the Dredging Contractors of America (DCA)—a national association for the dredging industry—to update industry cost data. Corps documentation related to the effort indicated that the Corps learned important information through discussions with industry, and a senior Corps cost-estimating official that we spoke with said that, on the basis of these discussions, the Corps updated the training it provides to Corps staff on preparing hopper dredge cost estimates.

Some of the data the Corps uses in preparing its hopper dredging cost estimates, however, remain outdated despite the Corps' attempt to update the information. Specifically, the Corps has not obtained updated technical data on industry hopper dredge equipment or labor rates but instead is relying on outdated information, some of which dates back to the late 1980s. During efforts to update the Corps' cost-estimating data in 2008, the Corps prepared a survey to collect industry dredge equipment information from the five dredging companies that owned hopper dredges. In cooperation with the Corps, DCA sent the survey to the companies. In the August 2008 letter accompanying the survey, the dredging association stated that "much of the cost basis the Corps uses for industry dredges is old data and limited due to lack of industry input" and noted that the Corps' ability to obtain the data would be mutually beneficial to the companies and the Corps. Among other things, data the survey sought to collect included costs of dredge acquisition, capital improvements, and certain types of repairs. Efforts to obtain these data were unsuccessful, however, due in part to industry's concerns about sharing business-sensitive data with the Corps. Industry representatives from one hopper dredging company we spoke with explained that they were concerned that cost data provided to the Corps might become accessible to their competitors and therefore the data were not provided. A senior Corps cost-estimating official we spoke with told us that the Corps limits the release of cost data used in preparing cost estimates within the Corps and that updated industry cost data would assist the Corps in preparing its cost estimates for hopper dredge work. The official also stated that other efforts could be made to obtain updated cost data, including performing a Corps-wide study to evaluate information from each Corps district office with hopper dredging contracts or reviewing contract audits. The Corps, however, has no plans for conducting such a study. In conducting a study, the Corps could assess the most effective and efficient approach for obtaining updated cost data, including examining whether and to what extent it would base its study approach on a review of contracts or contract audits, working directly with industry, or other approaches. Federal internal control standards state the need for federal agencies to establish plans to help ensure goals and objectives can be met.[15] A written plan would assist the Corps in obtaining updated cost data and following sound cost estimating practices, as described in our 2009 cost estimating and assessment guide, which is a compilation of cost-estimating best practices drawn from across government and industry.[16] Obtaining reliable and up-to-date data are important for developing sound cost estimates, and the Corps' cost estimate credibility may suffer if technical data are not updated and maintained, as noted in our cost estimating guide.

The Corps Analyzed Options for Operating Its Hopper Dredges in a Report to Congress

In response to our 2003 recommendation that the Corps prepare a comprehensive analysis of the costs and benefits of existing and proposed restrictions on the use of the Corps' hopper dredge fleet, the Corps prepared an analysis of its fleet for a 2005 report to Congress.[17] In its report, the Corps analyzed a number of options for operating its hopper dredges and made a recommendation to Congress for adjusting its fleet based on costs and benefits outlined in its analysis. The Corps recommended an option that it said would, among other things, ensure there was a viable reserve capability ready to respond to unforeseen requirements and ensure the timely accomplishment and reasonable cost for federal projects requiring hopper dredges. Under the option it recommended, the Corps would have (1) increased the *Essayons's* dredging by about 35 days, and kept the *Yaquina's* dredging days the same; (2) continued to keep the *Wheeler* in ready reserve; and (3) retired the *McFarland*. The Water Resources Development Act of 2007 did not specifically address these recommendations, but instead placed the McFarland in ready reserve and removed the then-existing restrictions on the *Essayons* and *Yaquina*.

Statutory Restrictions on Corps Hopper Dredges Have Resulted in Additional Costs to the Corps, but Effects on Industry Competition Are Unclear

Since 2003, statutory restrictions on the use of the Corps' hopper dredges have resulted in additional costs, but it is unclear whether the restrictions have affected competition in the hopper dredging industry.[18] Restrictions effectively limiting the number of days that Corps dredges can work have resulted in additional costs to the Corps, such as costs to maintain the ready reserve vessels while idle. On the other hand, the restrictions help ensure the Corps' ability to respond to urgent and emergency dredging needs when industry dredges may be unavailable. The extent to which restrictions on the use of the Corps' hopper dredges have affected competition in the dredging industry—as measured by the number of companies with hopper dredges and the number of bidders and winning bid prices for Corps projects—is unclear, based on our analysis of data on industry bids per Corps solicitation and other factors.

Statutory Restrictions Have Resulted in Additional Costs for Corps Hopper Dredging

Since 2003, statutory restrictions on the use of the four Corps' hopper dredges—in particular, the *Wheeler* and the *McFarland*—have resulted in additional costs to the Corps. First, the vessels have needed annual funding to maintain them in ready reserve because, given their limited use, the Corps is unable to recoup their costs with revenues from dredging work. The Corps incurs many of the costs for its hopper dredges—such as paying a crew and keeping engines and other systems in ready working condition—regardless of how much the dredges are used. For instance, placing the *McFarland* in ready reserve resulted in a substantial decrease in its dredging work (as measured in days worked and amount of material removed) but a relatively small decrease in its operating costs. As shown in table 2, the average annual cubic yards of material removed by the *McFarland* declined by 60 percent, while its average annual operating costs declined by 16 percent.[19]

Annual funding needed to maintain the *Wheeler* and the *McFarland* in ready reserve, which is provided through the Corps' civil works appropriation, has increased since 2003. Specifically, in fiscal year 2003, ready reserve funding for the *Wheeler* was $7.6 million, and it increased to $13.6 million in fiscal year 2012. In addition, the *McFarland* has received ready reserve funding of over $11 million each fiscal year since it was placed in ready reserve, resulting in total ready reserve funding for the vessels of over $25 million in fiscal year 2012 (see figure 3).

Second, the ready reserve restrictions have contributed to increases in the daily rate the Corps charges projects for use of the *Wheeler's* service, and future increases in the *McFarland's* daily rate may also be needed if it experiences unanticipated cost increases.[20] Increases in daily rates may result in either increasing costs, fewer cubic yards of material removed, or both, for the projects that use the *Wheeler* and *McFarland*—primarily projects in the Delaware River and the Mississippi River mouth, respectively. Officials from Corps headquarters and district offices responsible for the ready reserve hopper dredges told us they set the dredges' daily rates in part based on how many days they expect the dredges to work in the coming year and that, in the case of the *Wheeler*, the limited dredging days since being placed in ready reserve have contributed to higher daily rates. For instance, the *Wheeler's* daily rate has increased from $75,000 in fiscal year 2003 to $140,000 in fiscal year 2012, and the Corps expects a rate of $165,000 during fiscal year 2014. Furthermore, although costs for industry hopper dredge work have also increased, officials from a Corps district office that historically used the *Wheeler* told us that they would now be reluctant to use the vessel instead of an industry hopper dredge because of its high daily rate. In the case of the *McFarland*, the Corps has increased the vessel's daily rate from $94,000 in fiscal year 2009 (the last full fiscal year before it was placed in ready reserve) to $100,000 in fiscal year 2012, and officials said they planned to increase and then maintain the daily rate at $110,000 for the next several fiscal years. If there are unanticipated increases in costs for the *McFarland*, however, such as an unexpected increase in repair costs, Corps officials said they would likely have to increase the vessel's daily rate to cover such costs. As the officials explained, they set the *McFarland's* daily rate with an expectation that the vessel will work 70 days because the ready reserve restrictions do not allow them to increase the number of days the *McFarland* can work. Therefore, raising the vessel's daily rate would be the Corps' primary option to cover an increase in costs.

Table 2. Summary of Operations and Cost Data for the Corps Dredge *McFarland*

Operations and cost data	Before ready reserve (fiscal years 2008 and 2009)[a]	After ready reserve (fiscal years 2011 and 2012)[a]	Percentage change
Average days worked	144	85	-41%
Average cubic yards of material removed	1,706,570	685,957	-60%
Average operating cost	$9,208,887	$7,766,939	-16%
Labor costs	$4,420,847	$3,919,780	-11%
Fuel, lubricants, and water costs	$2,545,462	$1,677,801	-34%
Other operating costs[b]	$2,242,578	$2,169,358	-3%

Source: GAO analysis of U.S. Army Corps of Engineers data.

Note: All cost amounts are in constant 2013 dollars. Hopper dredges in ready reserve are used primarily for training purposes and can be called into active status only in limited circumstances, such as emergencies.

[a] The *McFarland* was placed in ready reserve on December 30, 2009. Therefore, the period before ready reserve includes the 2 full fiscal years prior to ready reserve—2008 and 2009—while the period after ready reserve includes the 2 full fiscal years following ready reserve—2011 and 2012. Limiting the period before ready reserve to 2 fiscal years reduced the effect of long-term changes, such as increasing fuel prices, on the *McFarland's* costs prior to ready reserve.

[b] Other operating costs include overhead, crew training, and information technology services, among others.

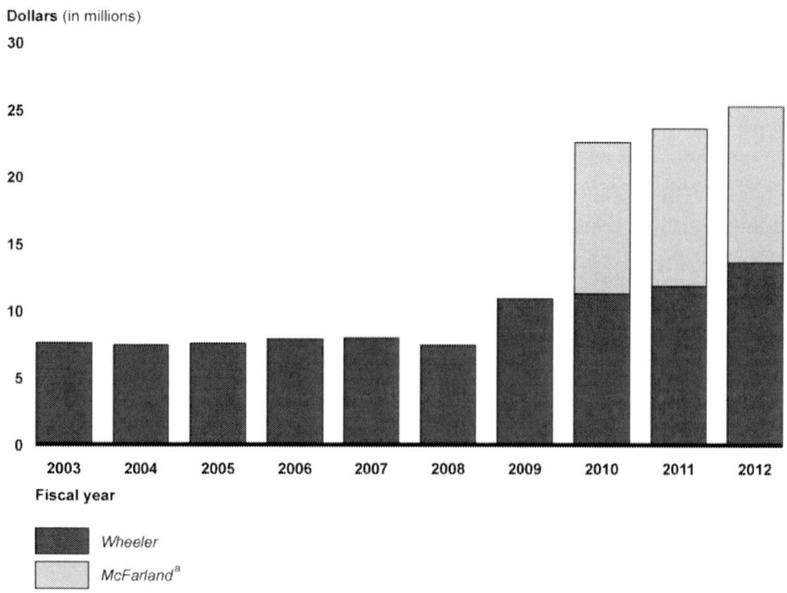

Source: U.S. Army Corps of Engineers.

Note: Hopper dredges in ready reserve are used primarily for training purposes and can be called into active status only in limited circumstances, such as emergencies.

[a] The *McFarland* was placed in ready reserve on December 30, 2009, and it began receiving ready reserve funding in fiscal year 2010.

Figure 3. Ready Reserve Funding for the Corps Dredges *Wheeler* and *McFarland*, Fiscal Years 2003 through 2012.

On the West Coast, restrictions on the number of days the Corps' hopper dredges *Essayons* and *Yaquina* could work had led to inefficiencies in completing their work before those restrictions were lifted by the Water

Resources Development Act of 2007, according to Corps officials. Before the 2007 act, the *Essayons* and the *Yaquina* were restricted to working about 180 workdays annually and, for several years, they reached their operating limits and, therefore, had to return to port before the projects they were working on were finished. The dredges were then sent back to complete the projects once the new fiscal year began, which was in October when weather conditions had begun to deteriorate. As a result, the Corps incurred additional transit and payroll costs while returning to complete the projects. Since the restrictions on these dredges were removed under the 2007 act, Corps officials said they have not had to interrupt ongoing work due to operating limits on the dredges and have had greater flexibility regarding when to perform work.

Ready Reserve Restrictions Help Ensure the Corps' Ability to Respond to Critical Dredging Needs

The ready reserve restrictions on the *Wheeler* and *McFarland* help ensure that they are available to the Corps for responding to urgent and emergency dredging needs, especially in the regions where the dredges are stationed. Demand for hopper dredging often varies substantially from year to year, and month to month, due in part to severe weather events such as hurricanes and floods, other events such as the Deepwater Horizon oil spill in 2010, or environmental restrictions that limit dredging work to certain months of the year. This variability has resulted in periods of high demand during which the Corps has used its ready reserve hopper dredges to respond to urgent or emergency dredging needs when industry hopper dredges were not available. As the Corps noted in its 2005 report to Congress, having the *Wheeler* in ready reserve is important to ensure that the vessel is available when unforeseen dredging needs occur, while more fully utilizing the *Wheeler* could limit the Corps' capability to respond to peak workload demands.[21] Specifically, the Corps has used the *Wheeler* to respond to urgent or emergency dredging needs 15 times during fiscal years 2003 through 2012.[22] In these cases, according to Corps documents, industry dredges were unavailable to immediately respond to time-sensitive dredging needs at the mouth of the Mississippi River, and the Corps was able to quickly move the *Wheeler* to the site and conduct the work. Similarly, local pilots and a local port authority we spoke with told us that the *McFarland* has been critical in addressing dredging needs on the Delaware River and Bay, where the vessel is stationed in ready reserve. Since its placement in ready reserve at the end of 2009, the Corps has used the

McFarland to respond to urgent or emergency needs 4 times. Industry representatives from most dredging companies we spoke with agreed that there is a need for Corps hopper dredges, specifically those placed in ready reserve, to respond to urgent or emergency situations when industry hopper dredges are unavailable.

The Extent to Which Restrictions Have Affected Industry Competition Is Unclear

Since 2003, the extent to which restrictions on the use of the Corps' hopper dredges have affected competition in the dredging industry—as measured by the number of companies with hopper dredges and the number of bidders and winning bid prices for Corps projects—is unclear. A possible benefit of restrictions on the amount of work performed by the Corps' hopper dredges is that the increased demand for industry hopper dredging services could encourage existing firms to add dredging capacity or new firms to enter the market, which could promote competition, raising the number of bidders and lowering winning bid prices for hopper dredging contracts. In addition, according to dredging industry representatives we spoke with, the more industry dredges can be utilized instead of Corps dredges, the lower the contract prices will be because contractors can spread their costs over more days of operation. However, on the basis of our analysis of (1) the dredging industry, (2) the number of bidders and bid prices for Corps dredging contracts, and (3) other factors that may have affected the level of competition for hopper dredging contracts, it is unclear whether or to what extent the restrictions on the Corps' hopper dredges may have increased the level of competition in the hopper dredging industry.

First, since 2003, the number of companies with hopper dredges in the United States has not changed, although the number of industry hopper dredges and the total size of these dredges have decreased. Specifically, at the end of 2013, five companies operated one or more hopper dredges. The same number of companies operated hopper dredges in 2003. Of the five companies we reported on in 2003, two sold their hopper dredges and exited the hopper dredging market while two new companies that had not been in the market acquired hopper dredges, and three companies remained the same. Since 2003, the total number of industry vessels decreased from 16 to 13, and the total capacity of these vessels, as measured in cubic yards, decreased by 16 percent. The decrease from 16 to 13 vessels resulted from one company relocating four of its U.S. hopper dredges overseas to perform dredging work primarily in the Middle East,[23] while another company built a new hopper dredge for the U.S. markct. In addition, as of January 2014, onc company had begun building a new hopper dredge that it expects will be completed in late 2014 or early 2015, and another company announced plans to build a new hopper dredge that it expects will be completed in 2015. If no companies remove existing hopper dredges from the U.S. market, these two dredges, if built as planned, would increase total industry capacity to 13 percent above 2003 levels. According to industry representatives with whom we spoke, dredging companies consider restrictions on the Corps' hopper dredges in deciding whether to acquire or build a new hopper dredge, but they also consider other factors, such as anticipated funding levels by the Corps, as well as nonfederal work.

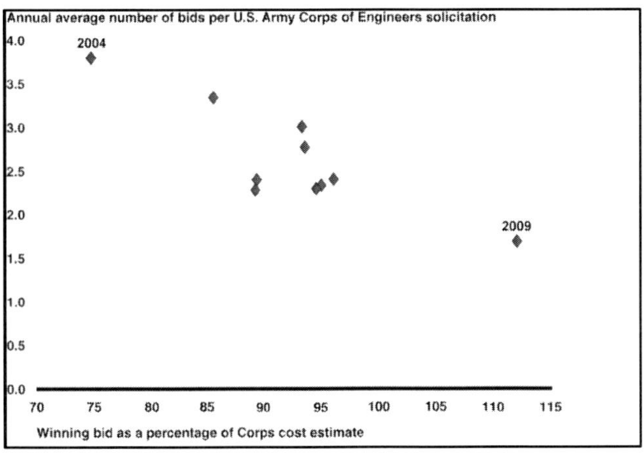

Source: GAO analysis of U.S. Army Corps of Engineers.

Note: Each point represents a fiscal year. We limited this analysis to awarded, sealed-bid solicitations because those are the solicitations for which the Corps has reliable data on numbers of bids and bid prices. We also limited the analysis to maintenance projects because, according to Corps officials and industry representatives, new construction projects are inherently uncertain and, therefore, new construction cost estimates and bid prices have greater variation and less accuracy than those for maintenance projects, which tend to occur regularly and have a history of price information.

Figure 4. Annual Average Number of Industry Bids per Corps Solicitation and Winning Bid as a Percentage of the Corps' Cost Estimate, Fiscal Years 2003 through 2012.

Table 3. Winning Bid Prices and Number of Bids for East Coast Maintenance Hopper Dredging Projects, Before and After the *McFarland's* Placement in Ready Reserve

	Before ready reserve (fiscal years 2003 through 2009)	After ready reserve (fiscal years 2011 and 2012)[a]
Winning bid as a percentage of Corps' cost estimate	84 percent	92 percent
Average number of bids per solicitation	2.9	2.8

Source: GAO analysis of U.S. Army Corps of Engineers data.

Note: We limited this analysis to awarded, sealed-bid solicitations because those are the solicitations for which the Corps has reliable data on numbers of bidders and bid prices. In sealed bidding, a contract is awarded to the responsible bidder with a bid that conforms to the invitation for bid and is the most advantageous for the government considering price and price-related factors included in the invitation. Approximately 76 percent of hopper dredging contracts awarded by the Corps from fiscal year 2003 through fiscal year 2012 were awarded through a sealed-bid process. We also limited the analysis to maintenance projects, because Corps dredges do not generally perform other types of projects such as new construction and, therefore, restrictions on Corps dredges would not have had an effect on industry competition for such projects.

[a] The period after ready reserve includes the 2 full fiscal years following the *McFarland's* placement in ready reserve on December 30, 2009.

Second, we did not find evidence of increased competition based on the number of bidders and winning bid prices for Corps hopper dredging projects since 2003. Economic principles suggest that an increase in the number of competitive bidders in the market should lead to lower prices. The correlation between the number of companies competing for hopper dredging contracts and the winning bid prices for those contracts is demonstrated by the Corps' historical data.[24] As shown in figure 4, in years where there were more industry bids per Corps solicitation, the average winning industry bid, as a percentage of the Corps' cost estimate, was generally lower, consistent with economic principles.[25]

Moreover, available Corps data related to the placement of the *McFarland* in ready reserve do not show evidence of increased competition in the dredging industry. Specifically, as shown in table 3, after the *McFarland* was placed in ready reserve, average winning bid prices increased for East Coast maintenance projects (i.e., projects the *McFarland* might undertake if use of the vessel were not restricted), and the average number of bids for those same projects decreased slightly.[26]

Third, other factors aside from the ready reserve restrictions may have affected the level of competition in the dredging industry since 2003. Examples of such factors include the following:

- *Environmental restrictions:* Multiple Corps officials and industry representatives told us that environmental restrictions related to endangered sea turtles and other species—which prohibit dredging during the time of year that those species are present—have contributed to fewer bidders for hopper dredging projects, particularly on parts of the East Coast. For instance, because of environmental restrictions, navigation dredging in fiscal year 2014 is limited to December 15, 2013, through March 31, 2014, in much of the Corps' South Atlantic Division,[27] during which time there are 48 potential Corps dredging projects planned, according to a 2013 Corps planning document. Corps officials attributed the absence of awardable bids for several recent East Coast hopper dredging solicitations to the unavailability of industry hopper dredges when the projects were scheduled to occur—during the period of high demand for hopper dredges caused by environmental restrictions. In addition, they expressed concern that similar shortages of bids could occur in the future.
- *Coordination among Corps district offices.* Increased coordination in scheduling hopper dredging projects across Corps district offices has helped distribute projects more evenly over time so that more companies had hopper dredges available with which to bid on projects, according to Corps officials. In contrast, when a large number of projects occur at the same time, dredging companies may not have enough dredges available to bid on all projects, thereby reducing the number of bidders for the projects. According to Corps officials we spoke with, increased regional coordination and sharing of up-to-date information on upcoming dredging needs across district offices has helped the Corps to better inform industry of planned work and align the scheduling of projects with the availability of industry dredges. In particular, Corps officials said increased coordination helped the Corps avoid scheduling too many projects simultaneously during a period of increased demand for hopper dredging work

following Hurricane Sandy and a Gulf Coast rebuilding effort to protect against the coastal impacts of oil spills.

- *Demand for nonfederal hopper dredging work.* Corps officials and industry representatives also told us that demand for hopper dredging work from states, private sources, and foreign governments has reduced the number of industry hopper dredges available for Corps projects. For instance, following the Deepwater Horizon oil spill in 2010, there was an increase in private and state funding for hopper dredge work to construct barrier islands to protect the coastline from the effects of the oil spill. Demand for hopper dredges for this work affected the dredges' availability for Corps navigation projects, according to Corps documents and officials, and industry representatives. In addition, representatives from one company said that, in part, because of increasing demand for hopper dredges from foreign governments—specifically in the Middle East—the company relocated several hopper dredges overseas, removing them from the U.S. market.
- *Differences in hopper dredge capabilities.* Because there are important variations in the size and capabilities of hopper dredges, the requirements of specific dredging projects can result in a limited number of dredges that may be able to effectively compete for a particular dredging project. For instance, the state of California requires hopper dredges to use reduced-emissions engines, in accordance with state air quality regulations. Of the 13 industry hopper dredges, only 3 have such engines, according to a Corps official. Similarly, according to Corps documents, a hopper dredge working at the mouth of the Columbia River in Oregon must be able to dredge against strong currents and endure large waves—capabilities that less than half of the industry fleet possesses, according to a Corps official. Other requirements, such as the depth of the waterway being dredged, or whether the material removed needs to be pumped onto the shore, can also limit which dredges can effectively compete for and carry out the work.

The Corps Faces Challenges Regarding the Fiscal Sustainability and the Future Composition of Its Hopper Dredge Fleet

Key challenges the Corps faces in managing its hopper dredge fleet are (1) ensuring the fiscal sustainability of its hopper dredges and (2) making decisions about the future of its hopper fleet composition, including the utilization of its existing fleet, changes to its existing fleet—including repairs, and the replacement or retirement of any vessels—and the utilization of any new replacement vessels.

The Corps Faces Challenges in Ensuring the Fiscal Sustainability of Its Hopper Dredges

The Corps faces challenges in ensuring the fiscal sustainability of its hopper dredges. In a 2012 study the Corps conducted on the fiscal condition of its hopper dredges, it identified increasing ownership and operating costs for its four hopper dredges, among other things, as a cause for concern and stated that the dredges would become unaffordable unless actions were taken.[28] For instance, the Corps' study projected that, in fiscal year 2012, the Corps' total end of fiscal year account balance for its four hopper dredges would exceed their funding levels by over $15 million dollars, and that fiscal problems would continue for the four hopper dredges through fiscal year 2016. The Corps stated in the study that it was concerned that project funding, which the Corps' hopper dredges depend on to varying degrees, was not increasing and, in some cases, was decreasing.

The Corps' 2012 study identified several actions to take to operate all of its hopper dredges with a positive account balance by the end of fiscal year 2015. For example, based on the study, a corresponding July 2012 implementation memorandum, and our discussions with Corps officials, the Corps

- increased the daily rates all four of the Corps' hopper dredges charge to projects that use the dredges, beginning in fiscal year 2012;
- increased funding in fiscal years 2013 and 2014 budgets for projects that use its hopper dredges to compensate for the vessels' corresponding increases in daily rates; and

- formed a team to conduct a hopper dredge operating cost review including, among other things, an evaluation of the affordability of two hopper dredges, the *Wheeler* and the *Yaquina*, by June 30, 2014.[29]

Corps officials told us that the actions identified in its study will help return the Corps' hopper dredges to a fiscally sustainable position by the end of fiscal year 2015. Since the Corps prepared its study, however, two of its hopper dredges encountered problems that significantly increased the dredges' total deficit: the *Essayons* had a dredging accident in 2013, and the *Wheeler* experienced an unforeseen delay in completing its engine replacement.[30] Even with these added costs, Corps officials said they expect to be able to reach their goal of returning each of the hopper dredges to fiscal sustainability by the end of fiscal year 2015, but they also acknowledged that this goal has become more challenging. For example, the officials said that, while the *Essayons'* accident increased its deficit by about $2 million, increasing the vessel's daily rate in fiscal year 2014 and dredging work in fiscal years 2014 and 2015 would give the vessel a positive account balance. Corps officials acknowledged that the *Wheeler's* situation was more precarious because it ended fiscal year 2013 with a deficit of over $5 million more than projected in the Corps' 2012 study, given the engine replacement delay. To get the *Wheeler* to a positive account balance by the end of fiscal year 2015, Corps officials said that they anticipated increasing the *Wheeler's* daily rate and potential dredging activity to more than 70 days under ready reserve in fiscal year 2014. Corps officials said they believe they have some flexibility with the number of days the vessel can dredge since there is not a set amount specified in statute.[31] Corps officials stated they are not planning further actions beyond those identified in the 2012 study at this time, but they acknowledged that additional measures, such as pursuing a permanent increase in the amount of days that the *Wheeler* may dredge each year under ready reserve, might be warranted if the vessel's fiscal situation does not improve by the end of fiscal year 2014.

The Corps Faces Challenges in Determining the Future Composition of Its Hopper Dredge Fleet

The Corps also faces challenges in making decisions about the future composition of its hopper dredge fleet. Some of the factors that make it difficult for the Corps to determine what composition of its fleet would best allow it to conduct dredging activities in the manner most economical and advantageous to the United States include the following:

- *Aging Corps' fleet.* The aging of the Corps' hopper fleet, contrasted with the millions of dollars the Corps has invested to upgrade the vessels, has made it challenging for the Corps to determine the long-term sustainability of its hopper dredges. Three of the Corps' four hopper dredges—*Essayons*, Wheeler, and *Yaquina*—have been in service for at least 30 years, and the *McFarland* has been in service over 45 years. According to Corps documentation, the Corps plans a 50-year investment life for its hopper dredges and, based on historical records, major repairs are typically needed when a dredge is about 30 years old. Since 2009, the Corps has invested millions of dollars in replacing and upgrading needed equipment on its four hopper dredges. For example, among other things, the *Essayons*, *Wheeler*, and *Yaquina* all had their engines replaced within the last 5 years allowing them to meet higher air quality emission standards. Similarly, the *McFarland's* electrical systems were replaced in fiscal year 2011, which increased the vessel's efficiency, since many of the systems were original equipment. According to Corps documents and officials, overall, all four hopper dredges are in good operating condition, but given the age of the vessels, the Corps has recognized the need to assess future repair or replacement options for its hopper dredges.
- *Effects on industry.* Because the Corps relies on both its own dredges and industry dredges to complete hopper dredging work, it needs to factor in both fleets in making future decisions about the composition of its own fleet. As of March 2014, 13 hopper dredges in the U.S. industry fleet had been in service for an average of about 27 years, though information on the extent to which these vessels have been maintained, upgraded, or may be close to going out of service has not been shared by industry with the Corps. During a discussion with industry representatives, however, representatives said that the hopper dredging industry is driven by competition, and they maintain their dredges to be as efficient as possible to improve their competitiveness in the market. Corps officials from several district offices we spoke with said that, because of the increasing use of industry hopper dredges for nonfederal beach nourishment projects, as well as anticipated increases in federal hopper dredging projects, industry's

availability to respond to the nation's navigation dredging needs may be stretched. These officials said that, as a result, maintaining the Corps' current fleet composition and perhaps increasing the use of some of the vessels, may be warranted. In contrast, most of the industry representatives we spoke with said they believe that industry has the ability to handle any increases in dredging projects, and the Corps' fleet should be further restricted or even reduced. These representatives stated that if the Corps increased its hopper dredge capability, then industry's portion of the overall dredging work would be reduced possibly leading companies to increase prices to cover their operating costs or potentially relocate their hopper dredges overseas.

- *Funding uncertainties.* Variability regarding federal funding for dredging also poses challenges to the Corps' plans for its fleet. While funding for hopper dredging has increased since fiscal year 2003 and was about $370 million in fiscal year 2012, Corps officials and stakeholders we spoke with said that, at recent funding levels, there were substantial unmet hopper dredging needs such as providing dredging for small ports and harbors.[32]

In a 2011 study related to the capital investment in its minimum dredge fleet, the Corps recognized the need to examine whether it should change the composition of its hopper fleet, such as by retiring one or more of its vessels, with or without a replacement.[33] Among the reasons the Corps listed for preparing this study were increases in the dredges' operating costs, the estimated costs of replacing the vessels, their increasing age, and the potential risk to navigation from reduced minimum fleet availability. The study found that the Corps' current approach to operating its minimum fleet would eventually increase the risk to navigating the waterways the Corps was responsible for dredging. To minimize that risk, the study and its corresponding August 2013 implementation memorandum, identified several actions, along with targeted time frames, that the Corps should take, including

- evaluating the *McFarland's* replacement options in 2017, which would consider replacing it and the *Wheeler* with one medium-sized hopper dredge;[34]
- deferring minimum fleet sustainment and improvement actions, such as modifications to improve certain hopper dredges' engine emissions, until at least fiscal year 2016 and possibly longer; and

- conducting a life-cycle cost analysis to support funding plans for future dredging needs which would include a cost comparison to either (1) use and then replace the vessels or (2) repair and sustain the vessels.

The 2011 study developed options based on three funding scenarios—increased, sustained, or decreased—and, as stated in the study and the Corps' implementation memorandum, the Corps selected the option associated with sustained funding levels as the best course of action. Should increased funding become available for dredging, a Corps official we spoke with said the Corps may need to adjust its planned course of action. The officials said that the 2011 study could provide the Corps with direction for adjusting its actions. For example, as noted in the study under the increased funding scenario, the Corps could continue with its planned fleet improvements instead of deferring them under the sustained option.

CONCLUSION

Hopper dredges play a vital role in keeping the nation's ports, harbors, and other waterways open for commerce. Over the past several decades, the Corps has increasingly relied on industry to carry out hopper dredging work, but it has also maintained its own minimum fleet of four hopper dredges, in part to ensure its ability to respond to critical dredging needs during periods of high demand. The Corps is faced with the task of balancing the hopper dredging work it contracts out to industry and maintaining the viability of its own fleet. The Corps has recognized the need to make changes to manage its hopper dredge fleet in a fiscally sustainable manner and has taken several actions to do so, including assessing the need to potentially modify the composition of its fleet. Since our 2003 report, the Corps has also made progress in addressing our recommendations to improve the information it maintains to manage its hopper dredging program, including modifying data fields in its dredging database to track solicitations that receive no bids or where all the bids received exceeded the Corps' cost estimate by more than 25 percent. However, because Corps district offices are not consistently populating the database with these solicitation data, the Corps does not have accurate or complete information that may help it identify potential gaps in industry's ability to fulfill certain dredging needs, which could inform its plans for future hopper dredging work. Additionally, the Corps made attempts to update the industry cost data it uses to prepare its cost estimates for hopper dredging contracts. Yet, some of the data it relies on remain outdated, and the Corps has no plans to update the information, such as through a Corps-wide study. Until the Corps has a plan for obtaining and then consistently updating reliable cost data, the Corps' ability to ensure the soundness of its cost estimates may suffer.

RECOMMENDATIONS FOR EXECUTIVE ACTION

We recommend that the Secretary of Defense direct the Corps of Engineers to take the following two actions:

To ensure the Corps of Engineers has the information it needs to analyze and make informed decisions regarding future hopper dredging work, provide written direction to its district offices on the importance of and need to accurately and consistently populate the data fields in its dredging database that track solicitations that receive no bids or where all the bids received exceeded the Corps' cost estimate by more than 25 percent.

To assist the Corps in preparing sound and credible cost estimates for soliciting bids for hopper dredge work by industry, develop a written plan for conducting a study to obtain and periodically update data on hopper dredging costs for its cost estimates, including reliable data on industry hopper dredge equipment and labor rates.

AGENCY AND THIRD-PARTY COMMENTS AND OUR EVALUATION

We provided a draft of this report to the Department of Defense and the Dredging Contractors of America (DCA) for review and comment. In its written comments, the Department of Defense concurred with our recommendations and stated that (1) the Corps will issue a letter to the district offices reinforcing the need to provide accurately and timely information in the Corps' dredging database, including information for solicitations that receive no bids or where all the bids received exceeded the Corps' cost estimate by more than 25 percent and (2) the Corps will develop a written plan as resources allow. The Corps also provided technical comments that we incorporated, as appropriate.

DCA provided written comments, which are summarized DCA neither agreed nor disagreed with our recommendations but disagreed with several statements in our report and raised objections to certain aspects of our scope and methodology. We disagree with DCA's comments as discussed below. Specifically, in its comments, DCA disagreed with our statement that a direct and valid comparison of work performed by industry to work performed by the Corps is not possible and stated that a third-party consultant performed an analysis of the Corps and industry hopper dredges performing similar work. According to DCA's comments, the industry hopper dredges can work for significantly less than Corps dredges. As we state in our report, we believe that a number of factors prohibit a direct and valid comparison of the Corps' and industry's costs of performing hopper dredge work, including limits to the number of days some Corps' dredges may operate and differences between dredging projects, such as the type of material dredged. In providing its estimates of cost savings for industry dredging, DCA did not provide information indicating how or whether it took such factors into account or to enable us to evaluate the reasonableness of its estimates.

DCA also questioned how, if one of the fundamental conclusions of our study is that the Corps has not made sufficient progress to improve the accuracy of its cost estimates, we could use those same government cost estimates to make industry competitiveness inferences. We concluded, however, that it is unclear whether statutory restrictions have affected competition in the hopper dredging industry. In reaching that conclusion, we analyzed a number of factors—including the number of companies with hopper dredges, the number of bidders and winning bid prices for Corps projects, and other factors such as environmental restrictions, the demand for nonfederal hopper dredging work, and differences in hopper dredge capabilities. We agree that obtaining reliable and up-to-date data are important for developing sound cost estimates, and our report recommends that the Corps develop a written plan for conducting a study to obtain and periodically update data on hopper dredging costs for its cost estimates.

DCA disagreed with our discussion on the capacity of the industry hopper dredge fleet, stating specifically that one industry dredge, the Long Island, should not have been included in our analysis because it had not been used for maintenance dredging and had not been used on a project for quite a few years. For our report, we did not limit our analysis to particular types of hopper dredging projects, such as maintenance projects, and we compared industry's total capacity today with what we reported in 2003, which we believe is a valid comparison. Moreover, in its comments on our 2003 report on hopper dredging, DCA included the *Long Island* in its list of industry dredges to support its point that industry hopper dredging capacity had increased in the decade leading up to 2003. As a result, we continue to believe it was appropriate to include the *Long Island* as a part of our analysis.

DCA stated our analysis of how the Corps' manages its hopper dredges was not comprehensive or objective and questioned why we did not examine options for retiring or further reducing the use of Corps' dredges. DCA suggested that such an examination should take place and would be in line with the congressional intent of increasing the use of private industry dredges. However, DCA quotes selectively from the main statute that governs the Corps' hopper dredging activities.[35] While those portions of the law read in isolation could suggest that the Corps should take further steps to privatize its hopper dredge work, other provisions of the same law either (1) give the Corps broad discretion to implement its hopper dredge responsibilities or (2) directly restrict the Corps' ability to reduce or eliminate Corps' dredges. It was not the purpose of our report to examine policy options for carrying out the Corps' hopper dredge work, including those not presently authorized under statute. We did examine and discuss actions the Corps has taken or plans to take in managing its hopper dredges, which include, among other things, conducting a hopper dredge operating cost review and evaluating retirement or replacement options.

Sincerely yours,

Anne-Marie Fennell
Director, Natural Resources and Environment

APPENDIX I: OBJECTIVES, SCOPE, AND METHODOLOGY

This report examines (1) the actions the Corps has taken to address our 2003 recommendations for improving the information needed to manage its hopper dredging program and develop cost estimates for industry contracts; (2) the effects since 2003, if any, of the statutory restrictions placed on the use of the Corps' hopper dredges; and (3) key challenges, if any, the Corps faces in managing its hopper dredge fleet.

To conduct our work, we reviewed Pub. L. No. 95-269, which established the Corps' minimum fleet, the Water Resources Development Acts of 1996 and 2007, and other laws, regulations, and Corps' policy and guidance governing the Corps' use of hopper dredges. We interviewed officials from Corps headquarters, division offices, and the 9 Corps district offices with the largest hopper dredging workload during fiscal year 2003 through fiscal year 2012 (out of a total of 17 district offices that contracted with industry for hopper dredging work during the time period): Galveston, Jacksonville, Mobile, New Orleans, New York, Philadelphia, Portland, San Francisco, and Seattle. We also visited the Corps' four hopper dredges and one industry hopper dredge for informational tours of these vessels to gain a better understanding of their physical characteristics and operations. We interviewed representatives from the national association for the dredging industry, the Dredging Contractors of America, and the five dredging companies that own and operate hopper dredges—Cashman Dredging, Dutra Group, Great Lakes Dredge & Dock Company, Manson Construction Co., and Weeks Marine, Inc. We also interviewed other stakeholders involved in hopper dredging, including a national pilots' association and a national port authority association, and local pilots' associations and port authorities from the areas where Corps hopper dredges are stationed—New Orleans, LA; Philadelphia, PA; and Portland, OR. We focused our review on the 10-year period between fiscal year 2003—when we conducted our previous review of the Corps' hopper dredges[36]—and fiscal year 2012—the most recent year for which Corps information on hopper dredging was readily available. In addition, we focused our review on the four hopper dredges in the Corps' minimum dredge fleet during the period of our review: the *Essayons*, *McFarland*, *Wheeler*, and *Yaquina*, and did not include other dredge types.[37]

To examine the actions the Corps has taken to address our 2003 recommendations for improving the information needed to manage its hopper dredging program and develop cost estimates for industry contracts, we reviewed the Corps' process for tracking and analyzing data on solicitations for industry hopper dredging work, both at district offices and Corps headquarters. We also reviewed several sources of information the Corps maintains on the use of its hopper dredge fleet, including information maintained by the district offices responsible for operating the Corps' four hopper dredges, and the tracking log the Corps maintains with information on the urgent and emergency work its ready reserve vessels undertake. In addition, we reviewed the Corps' 2005 report to Congress on hopper dredges,[38] other Corps studies and policy documents on the use of its hopper dredge fleet, and the information it uses to prepare hopper dredging cost estimates when soliciting bids by industry, including the engineering regulations and the computer program used to develop government cost estimates. We interviewed officials from the Corps' cost engineering center of expertise, who were responsible for developing cost estimating policies and software and officials from Corps district offices who were responsible for developing cost estimates for hopper dredging projects.

To examine the effects since 2003, if any, of the statutory restrictions placed on the use of the Corps' hopper dredges, we reviewed the statutes, regulations, and Corps' reports and other documents governing the use of the Corps' hopper dredge fleet. To assess changes in industry hopper dredging contracts since 2003, we reviewed and analyzed data from the Corps' Dredging Information System, the database it uses to maintain information on each of its dredging projects, including data on the type and location of the dredging work, the type of contract, and the numbers of bids and bid prices for the contracts. To assess the reliability of the data, we interviewed officials from the Corps' Navigation Data Center who maintain the database, as well as officials from nine Corps district offices who are responsible for entering and updating data on their district offices' dredging activities. We reviewed documentation related to the database, such as the user's guide and data dictionary, and electronically tested the data for missing or erroneous values and, in several cases, obtained updated or corrected data from the Corps. We determined the data we used on the type and location of the dredging work, the type of contract, and the number of industry bids and bid prices for sealed-bid solicitations were sufficiently reliable for our purposes. We also analyzed financial data on the Corps' hopper dredges, including their operating and ownership costs, and income from ready reserve funding. To assess the reliability of the Corps' financial data, we interviewed Corps officials who maintain these data, compared the data to other sources of information on the Corps' hopper dredges, and obtained clarifying information from the Corps for certain items such as ready reserve funding. We determined the data were sufficiently reliable for our purposes. We obtained and reviewed information from the five dredging companies that own and operate hopper dredges, including information on their hopper dredges' capabilities, dredging work they performed, and changes to their hopper dredge fleet since 2003. We did not directly compare work performed by industry hopper dredges with work performed by the Corps' hopper dredges because, as we first reported in 2003, a direct and valid comparison of the Corps' and industry's costs to perform hopper dredge work is not possible due to various factors.[39]

To examine key challenges, if any, the Corps faces in managing its hopper dredge fleet, we reviewed Corps reports and financial data on its hopper dredge fleet, including a 2011 study on capital investment plans for its minimum fleet composition[40] and a 2012 study of the fiscal condition of its hopper dredges. In addition to reviewing the 2012 fiscal study, we also obtained and analyzed additional data related to the financial condition of the Corps' hopper dredges. We also obtained and reviewed the Corps' 2012 and 2013 implementation memorandums related to both studies and discussed with Corps officials the actions the Corps has taken—and plans to take—related to the memorandums. We examined changes and potential challenges the Corps faces related to managing its hopper dredge fleet, including dredging accidents, repair delays, and potential funding changes. We discussed general Corps fleet management and composition options with industry officials and the other stakeholders we interviewed.

We conducted this performance audit from January 2013 to April 2014 in accordance with generally accepted government auditing standards. Those standards require that we plan and perform the audit to obtain sufficient, appropriate evidence to provide a reasonable basis for our findings and conclusions based on our audit objectives. We believe that the evidence obtained provides a reasonable basis for our findings and conclusions based on our audit objectives.

APPENDIX II: THE U.S. HOPPER DREDGE FLEET

As of March 2014, 17 hopper dredges were operating in the United States, 13 of which were owned by industry (see table 4). In addition, 2 industry hopper dredges are expected to be added to the U.S. fleet by 2015.

Table 4. Corps and Industry Hopper Dredge Fleets, 2014

Owner	Vessel	Size[a]	Capacity (in cubic yards)[b]	Year built	Loaded draft(feet)
Cashman Dredging	*Atchafalaya*	Small	1,300	1980	14.7
Dutra Group	*Stuyvesant*	Large	9,870	1981	35.0
	Columbia	Medium	4,350	1986[c]	16.5
Great Lakes Dredge & Dock Company	*Liberty Island*	Large	6,540	2002	28.3

Table 4. (Continued)

Owner	Vessel	Size[a]	Capacity (in cubic yards)[b]	Year built	Loaded draft(feet)
	Terrapin Island[d]	Large	6,400	1981	22.3
	Dodge Island	Medium	3,600	1980	19.5
	Padre Island	Medium	3,600	1981	19.6
Manson Construction Co.	*Glenn Edwards*	Large	13,500	2006	28.0
	Bayport	Medium	4,855	1999	22.0
	Newport	Medium	4,000	1983	19.0
	Westport	Small	1,800	1978	11.0
Weeks Marine, Inc.	*R.N. Weeks*	Medium	4,000	1987	19.6
	B.E. Lindholm	Medium	4,000	1985	22.3
Corps	*Wheeler*	Large	8,256	1982	29.5
	Essayons	Large	6,852	1983	33.0
	McFarland	Medium	3,140	1967	23.1
	Yaquina	Small	1,042	1981	14.7
Hopper dredges planned					
Weeks Marine, Inc.	*Magdalen*	Large	8,500	Under construction; estimated completion late 2014 or early 2015	25.3
Great Lakes Dredge & Dock Company	New build	Large	15,000	Construction scheduled to begin in 2014; estimated completion 2015	29.0

Sources: U.S. Army Corps of Engineers and dredging industry.

[a] A hopper dredge's size is determined by the capacity of its hopper—small hopper dredges have a capacity of up to 3,000 cubic yards, medium hopper dredges have a capacity of 3,001 to 6,000 cubic yards, and large hopper dredges have a capacity over 6,000 cubic yards.

[b] The capacities of several Corps and industry hopper dredges listed differ slightly from their capacities listed in our 2003 report. The capacities listed in this table are based on current information provided by the Corps and dredging companies.

[c] The *Columbia* was originally built in 1944 to transport military equipment in World War II, was later converted to a hopper dredge, and began its service as a hopper dredge in 1986, according to Corps documents.

[d] The *Terrapin Island* was formerly the *Eagle I*, owned by Bean Stuyvesant LLC. Great Lakes Dredge & Dock Company acquired and renamed the dredge in 2007.

End Notes

[1] Pub. L. No. 95-269 (1978).

[2] Corps of Engineers, *Minimum Fleet Capital Investment Report, 2012-2061* (Washington, D.C.: Dec. 12, 2011; revised Apr. 26, 2013).

[3] The Corps began implementing the restrictions on the workdays for the *Essayons*, *McFarland*, and *Yaquina* on October 1, 1993, and placed the *Wheeler* into ready reserve on October 1, 1997. See table 1 for the statutory restrictions in place on the use of the Corps' hopper dredges, and how they have changed since fiscal year 2003.

[4] H.R. Rep. No. 104-695 at 158-59 (1996).

[5] Pub. L. No. 110–114, § 2047 (2007). The Corps placed the *McFarland* into ready reserve on December 30, 2009.

[6] GAO, *Corps of Engineers: Effects of Restrictions on Corps' Hopper Dredges Should Be Comprehensively Analyzed*, GAO-03-382 (Washington, D.C.: Mar. 31, 2003).

[7] Corps, 2011 *Minimum Fleet Capital Investment Report*.

[8] A hopper dredge's size is determined by the capacity of its containment area or its "hopper"—small hopper dredges have a capacity of up to 3,000 cubic yards, medium hopper dredges have a capacity of 3,001 to 6,000 cubic yards, and large hopper dredges have a capacity over 6,000 cubic yards.

[9] The 2007 act did not specify any change in the restrictions on the Wheeler; however, beginning in fiscal year 2008, the Corps increased the *Wheeler's* training days from 55 to 70 days. According to Corps officials, the Corps set the *Wheeler's* training days at 70 days to be consistent with the day limit established for the *McFarland* by the 2007 act.

[10] There have been other legislative attempts to further alter the operations of the Corps fleet. For example, a House-passed bill in 2006 would have eliminated all funding for operating and maintaining the *McFarland*, but not for decommissioning the vessel. H.R. 5427, § 104 (2006). A Senate appropriations bill for fiscal year 2012 would have required the Corps to fully utilize all Corps hopper dredges without restriction. H.R. 2354, § 110 (2011).

[11] In addition to being no more than 25 percent above the government cost estimate, bids must be considered responsive in order to be awarded. One element the contractor generally must have in order for the Corps to consider a bid to be responsive is access to the required equipment for performing the scope of work. In addition to sealed-bid solicitations, the Corps awards hopper dredging contracts using requests for proposals, which allow the Corps to evaluate proposals on the basis of criteria other than price. Using the request for proposals process, the Corps may award a contract on a sole-source basis after negotiating with a single company because, for instance, there is an urgent need for dredging and soliciting work from multiple contractors would cause unacceptable delays. Additionally, a bidder who provides a responsive bid must be considered a responsible bidder which entails, among other things, that the bidder has adequate financial capability.

[12] GAO-03-382.

[13] GAO, *Standards for Internal Control in the Federal Government*, GAO/AIMD-00-21.3.1 (Washington, D.C.: Nov. 1, 1999).

[14] As part of this recommendation, we also recommended that the Corps examine the policies related to calculating transit costs—the costs for an industry dredge to travel to a project site—because the Corps was relying on an expired policy for some of its solicitations. The Corps updated the regulation containing its policies on calculating transit costs in 2008.

[15] GAO/AIMD-00-21.3.1.

[16] GAO, *Cost Estimating and Assessment Guide: Best Practices for Developing and Managing Capital Program Costs*, GAO-09-3SP (Washington, D.C.: March 2009). The guide states that updating and keeping cost estimates current to reflect changes in technical assumptions is one of the features of a high-quality, cost-estimating process and a best practice that helps develop reliable cost estimates that management can use for informed decisions.

[17] U.S. Army Corps of Engineers, *Report to Congress: Hopper Dredges* (Washington, D.C.: June 3, 2005). The 2004 Energy and Water Appropriations Act conference report referenced our 2003 report on the Corps' hopper dredges and also directed, among other things, that the Corps prepare an analysis of the costs and benefits of the existing and proposed restrictions on the use of its hopper dredges.

[18] Similarly, we found, in 2003, that the restrictions on the Corps' hopper dredge fleet at that time had imposed costs on the Corps' dredging program, but had not yet resulted in proven benefits—such as more bids per Corps solicitation or lower prices for winning bids. See GAO-03-382.

[19] We found, in 2003, that in the 4 years after the *Wheeler* was placed in ready reserve, the average cubic yards it dredged per year decreased by 56 percent from its average over the 4 years prior to being placed in ready reserve, while its average costs decreased by 20 percent over the same period. See GAO-03-382.

[20] Daily rates for Corps hopper dredges have increased and may continue to increase due to several factors such as increasing fuel costs and changes in Corps accounting methods, in addition to ready reserve restrictions. While future increases in the *McFarland's* daily rate may be needed if unanticipated cost increases are experienced, such increases could also apply to any of the Corps' hopper dredges. We did not quantify the extent to which individual factors contributed to increases in daily rates.

[21] U.S. Army Corps of Engineers, *Report to Congress: Hopper Dredges* (Washington, D.C.: June 3, 2005).

[22] Private industry hopper dredges may also respond to urgent or emergency dredging needs, precluding the need for the Corps to use one of its dredges. The Corps does not systematically track data on when it contracts with private industry to respond to urgent or emergency dredging events, however, and so data are not available to compare the use of Corps' dredges and industry dredges to respond to urgent or emergency events.

[23] According to a representative from DCA, a lack of work in the United States was also a factor in the relocation of some of these industry vessels. A vessel may engage in dredging in U. S. waters only if it is built, owned, and flagged in the United States. As a result, vessels that are relocated overseas and operated under a foreign flag are unable to dredge in the United States. The process of reflagging under the U.S. flag is generally so costly that it would effectively cause a dredge to be unable to compete in the United States, according to dredging industry representatives. As a result, the Corps and the dredging industry do not consider U.S.-owned dredges operating in foreign markets in their planning or analyses of the U.S. market.

[24] This analysis shows a relationship between the number of bids per Corps solicitation and the winning bid prices for those solicitations, but it does not establish a cause-and-effect relationship.

[25] Similarly, we found, in 2003, that from fiscal year 1990 through fiscal year 2002, years with more industry bids per Corps solicitation for hopper dredging generally had lower winning bid prices on average relative to the Corps' cost estimate. See GAO-03-382.

[26] Similarly, we found, in 2003, that after restrictions were placed on the number of days the Corps' hopper dredges could work, the average number of bids per Corps solicitation decreased, and the percentage of winning bids below the Corps' cost estimate decreased.

[27] The Corps' South Atlantic Division extends from North Carolina to the eastern portion of the Gulf Coast.

[28] Army Corps of Engineers, *USACE Dredge Back to Black Plan* (Washington, D.C.: June 29, 2012).

[29] According to Corps criteria for these studies, the Corps' evaluation of these vessels will include examining current and future Corps mission requirements, and the affordability of the vessel, including operating costs, program support, likely alternative funding sources, planned investments along with those to date, and the additional hidden costs of vessel support (such as spare parts inventory, shoreside personnel and facilities, and the impacts of any personnel action, etc.).

[30] A Corps official told us that the August 2013 grounding accident that the *Essayons'* experienced while dredging made the vessel inoperable for about a month while it underwent repairs. In the case of the *Wheeler*, a Corps official estimated that the delays in replacing the *Wheeler's* engines caused the vessel to remain out of operation at least 4 months more than the Corps initially planned. In addition, during this time, a cruise vessel broke free from its moorings during a storm and collided with the *Wheeler* when it was in the repair yard, which further delayed the *Wheeler's* return to work, according to the Corps.

[31] The *Wheeler* is not statutorily limited to a specific number of annual training days under ready reserve, but, since fiscal year 2008, the Corps has targeted the vessel's work to 70 training days annually, which is the same number of days the *McFarland* is permitted to work by statute under ready reserve. Both ready reserve vessels can also be used in urgent or emergency situations beyond their 70-day limits. See table 1 for the statutory restrictions in place on the use of all the Corps' hopper dredges.

[32] As of the time of this report, pending water resources development legislation in both the House and Senate would propose increasing annual funding for maintenance dredging through fiscal year 2020 and beyond. In addition, the Senate bill would require the Corps to study, among other things, "the needs of the United States for dredging, including the need to increase the size of private and Corps of Engineers dredging fleets to meet demands for additional construction or maintenance dredging needed as of the date of enactment of this Act and in the subsequent 20 years." H.R. 3080 (as amended by the Senate) § 2024(b)(2) (2013).

[33] Corps, 2011 *Minimum Fleet Capital Investment Report*. This study encompassed all 10 dredges in the Corps' minimum dredge fleet, which includes the 4 hopper dredges reviewed in this report, and 6 other dredges of different types that are generally used for different dredging projects.

[34] According to the minimum fleet study, "replacement options" include replacement, recapitalization, retirement, or divestment. The study noted that the *McFarland* and *Wheeler* were two primary vessels where consolidation is most likely to be considered, although the study also showed that the *Wheeler* was not scheduled to be evaluated for replacement until 2037.

[35] 33 U.S.C. § 622.

[36] GAO-03-382.

[37] We did not include the Corps' two "special purpose" hopper dredges—the *Currituck* and the *Murden*—because though these two vessels share some characteristics with the four larger hopper dredges, such as having hoppers and drag arms, they are substantially smaller than the rest of the Corps' or industry hopper dredge fleet. Specifically, the *Currituck* has a hopper capacity of 315 cubic yards, and the *Murden* has a hopper capacity of 512 cubic yards.

[38] U.S. Army Corps of Engineers, *Report to Congress: Hopper Dredges* (Washington, D.C.: June 3, 2005).

[39] GAO-03-382. These factors include, among other things, design features in the Corps' vessels in support of national defense missions, which add weight to the vessels and make them less efficient than industry vessels; limits to the number of days some of the Corps' vessels may operate; and differences between dredging projects—such as type of material dredged, type of work and corresponding risk level, and distance from the dredging operations to the placement site.

[40] Army Corps of Engineers, *Minimum Fleet Capital Investment Report, 2012-2061* (Washington, D.C.: Dec. 12, 2011; revised Apr. 26, 2013) and Army Corps of Engineers, *USACE Dredge Back to Black Plan* (Washington, D.C.: June 29, 2012).

In: Hopper Dredging of Waterways
Editor: Eleanor G. Chance
ISBN: 978-1-61122-357-6

Chapter 2

CORPS OF ENGINEERS: EFFECTS OF RESTRICTIONS ON CORPS' HOPPER DREDGES SHOULD BE COMPREHENSIVELY ANALYZED*

United States Government Accountability Office

WHY GAO DID THIS STUDY

The fiscal year 2002 Conference Report for the Energy and Water Development Appropriations Act directed GAO to study the benefits and effects of the U.S. Army Corps of Engineers' (Corps) dredge fleet. GAO examined the characteristics and changing roles of the Corps and industry in hopper dredging; the effect of current restrictions on the Corps' hopper dredge fleet; and whether existing and proposed restrictions on the fleet, including the proposal to place the *McFarland* in ready reserve, are justified. In addition, GAO identified concerns related to the government cost estimates the Corps prepares to determine the reasonableness of industry bids.

* This is an edited, reformatted and augmented version of the United States Government Accountability Office publication, GAO-03-382, dated March 2003.

Why GAO Recommends

GAO recommends that the Secretary of the Army direct the Corps of Engineers to (1) obtain and analyze baseline data to determine the appropriate use of the Corps' hopper dredge fleet, (2) prepare a comprehensive analysis of the costs and benefits of existing and proposed restrictions on the use of the Corps' hopper dredge fleet, and (3) assess the data and procedures used to prepare the government cost estimate. The Department of the Army agreed with GAO's recommendations. The Dredging Contractors of America generally agreed with GAO's recommendations, but strongly disagreed that restrictions on the Corps' hopper dredges have not resulted in proven benefits.

What GAO Found

In response to 1978 legislation that encouraged private industry participation in dredging, the Corps gradually reduced its hopper dredge fleet from 14 to 4 vessels (the *Wheeler*, the *McFarland*, the *Essayons*, and the *Yaquina*) while a private hopper dredging industry of five firms and 16 vessels has emerged. Dredging stakeholders generally agreed that the Corps needs to retain at least a small hopper dredge fleet to (1) provide additional dredging capacity during peak demand years, (2) meet emergency dredging needs, and (3) provide an alternative work option when industry provides no bids or when its bids exceed the government cost estimate by more than 25 percent. In reviewing the cost estimation process, GAO found that the Corps' estimates are based on some outdated contractor cost information and an expired policy for calculating transit costs.

The restrictions on the use of the Corps' hopper dredge fleet that began in fiscal year 1993 have imposed costs on the Corps' dredging program, but have thus far not resulted in proven benefits. The Corps estimates that it spends $12.5 million annually to maintain the *Wheeler* in ready reserve, defined as 55 workdays plus emergencies, of which about $8.4 million is needed to cover the costs incurred when the vessel is idle. A possible benefit of restrictions on the Corps' vessels is that they could eventually encourage existing firms to add dredging capacity or more firms to enter the market, which, in turn, may promote competition, improve dredging efficiency, and lower prices. Although there has been an increase in the number of private industry hopper dredges since the restrictions were first imposed, the number of private firms in the hopper dredging market has decreased. In addition, during the same time period, the number of contractor bids per Corps solicitation has decreased, while the number of winning bids exceeding the Corps' cost estimates has increased.

Although the Corps proposed that the *McFarland* be placed in ready reserve, it has not conducted an analysis to establish that this action would be in the government's best interest. Specifically, in a June 2000 report to the Congress, the Corps stated that the placement of the *Wheeler* in ready reserve had been a success and proposed that the *McFarland* also be placed in ready reserve. However, when asked, the Corps could not provide any supporting documentation for its report. Furthermore, according to the Corps, future use of the *McFarland* will require at least a $25 million capital investment to ensure its safety, operational reliability, and effectiveness. Such an investment in a vessel that would be placed in ready reserve and receive only minimal use is questionable.

ABBREVIATIONS

Army	Department of the Army
Corps	U.S. Army Corps of Engineers
DCA	Dredging Contractors of America
GAO	General Accounting Office

March 31, 2003

The Honorable Ted Stevens
Chairman
The Honorable Robert C. Byrd
Ranking Minority Member
Committee on Appropriations
United States Senate

The Honorable James M. Inhofe
Chairman
The Honorable James M. Jeffords
Ranking Minority Member
Committee on Environment and Public Works
United States Senate

The Honorable C.W. Bill Young
Chairman
The Honorable David R. Obey
Ranking Minority Member
Committee on Appropriations
House of Representatives

The Honorable Don Young
Chairman
The Honorable James L. Oberstar
Ranking Democratic Member
Committee on Transportation and Infrastructure
House of Representatives

Keeping the nation's navigation channels and ports fully functioning is vital to U.S. commerce both domestically and abroad. In 2001, shipping vessels moved more than $700 billion of import and export cargo through the nation's ports and harbors. Vessels called dredges remove sediment from the bottom of navigation channels, ports, and harbors to maintain waterways at the navigable depths and widths necessary for shipping. There is a variety of dredge types, each designed to perform optimally under specific conditions. One dredge type—the hopper dredge—performs much of the dredging needed in ports, harbors, and access channels exposed to the ocean, where traffic and operating conditions render the use of other dredges inefficient or impractical. A hopper dredge pumps material (e.g., sand and water slurry) into its hopper where it is stored before being transported to the disposal site. (See figure 1.)

The U.S. Army Corps of Engineers (Corps) is responsible for dredging U.S. ports and harbors. The Corps is to carry out projects for improvements of rivers and harbors, by contract or otherwise, in the manner most economical and advantageous to the United States. Until 1978, the Corps performed all hopper dredging work with its 14 hopper dredges. In 1978, legislation directed the Corps to (1) contract out much of its dredging work to private industry as industry demonstrated that it could perform the work at reasonable prices and in a timely manner, (2) reduce the federal fleet as industry demonstrated its ability to perform, and (3) maintain a minimum fleet of federal vessels for emergency and national defense purposes. The act also directed the Corps to retain as much of the fleet as it determined necessary to ensure that the federal and private fleets together could carry out necessary dredging projects. As a result, the Corps reduced the size of its hopper dredge fleet to four vessels—

the *Wheeler,* the *McFarland,* the *Essayons,* and the *Yaquina.* In the 1990s, in an effort to further promote private industry participation in hopper dredging, the Congress imposed restrictions on the Corps' hopper dredge fleet. These restrictions (1) effectively reduced the annual work schedule of each of the Corps' hopper dredges from about 230 to about 180 workdays and (2) limited the *Wheeler* to 55 workdays per year plus emergencies (referred to as "ready reserve" status). Furthermore, in fiscal year 2002, the Congress directed the Corps to confine the *McFarland* to emergency work and operations in the Delaware River. The Corps was to periodically evaluate the effects of the ready reserve program on the costs, responsiveness, and capacity of the Corps' and private industry's hopper dredges.

The fiscal year 2002 Conference Report for the Energy and Water Development Appropriations Act directed GAO to study the benefits and effects of the Corps' dredge fleet. In response, we examined (1) the changing roles of the Corps and industry in hopper dredging and the characteristics of the hopper dredging industry; (2) the effect of restrictions currently in place on the Corps' hopper dredge fleet; and (3) whether the existing and proposed restrictions on the Corps' hopper dredges, including placing the dredge *McFarland* in ready reserve, are justified. In addition, during the course of our work we identified concerns related to the government cost estimates that the Corps prepares to determine whether industry bids for dredging work are reasonable.

Source: U.S. Army Corps of Engineers.

Figure 1. Hopper Dredge.

RESULTS IN BRIEF

In accordance with legislative direction, the Corps has reduced its hopper dredge fleet, while the private hopper dredging industry has steadily increased its share of the annual hopper dredging workload. Today, of the 20 hopper dredges in service in the United States, five private industry firms operate 16 vessels and perform about 72 percent of the nation's hopper dredging maintenance work, while the Corps operates 4 vessels and performs the remaining 28 percent of the work. Corps officials and representatives from the dredging industry, selected ports, and the maritime industry generally agreed that the Corps needs to retain a hopper dredge fleet to (1) provide additional dredging capacity during peak demand years, (2) meet the emergency and national defense needs identified in the 1978 legislation, and (3) provide an alternative work option at times when industry offers unreasonable bids or no bids at all. To determine the reasonableness of private contractor bids, the Corps develops a government cost estimate that serves as a benchmark against which industry bids are compared. If the bids exceed the government estimate by more than 25 percent, the Corps may elect to perform the work itself.

During our review, we identified a number of concerns regarding the Corps' government cost estimate. Specifically, the Corps uses outdated industry cost data to determine the reasonableness of contractor bids when developing its cost estimate. In addition, the Corps continues to follow a policy that expired in 1994 when calculating contractor transit costs to the dredging site for some of its contracts. These concerns raise questions about the practical value of using the Corps' cost estimate as protection against high industry bids.

Restrictions on the use of the Corps' hopper dredge fleet, which began in fiscal year 1993, have imposed costs on the Corps' dredging program, but have thus far not resulted in proven benefits. Most of the costs of the Corps' hopper dredges are incurred regardless of how frequently the dredges are used. For example, the Corps' placement of the *Wheeler* in ready reserve—55 annual workdays plus emergencies—reduced the vessel's productivity by 56 percent but reduced costs by only 20 percent. The Corps estimates that it spends $12.5 million annually to maintain the *Wheeler* in ready reserve, of which approximately $8.4 million is needed to cover the costs incurred when the vessel is idle. A possible benefit of restrictions on the Corps' vessels is that they could eventually encourage existing firms to add dredging capacity or more firms to enter the market, which, in turn, may promote more competition, improve dredging efficiency, and lower prices. Although there has been an increase in the number of private industry hopper dredges since the restrictions were first imposed, the number of private firms in the hopper dredging market has decreased. In addition, during the same time period, the number of contractor bids per Corps solicitation has decreased, while the number of winning bids exceeding the Corps' cost estimate has increased. Another potential benefit of the restrictions is enhanced Corps responsiveness to emergency dredging needs. However, the Corps is unable to evaluate whether emergency dredging needs have been met more or less efficiently since the restrictions went into effect because it does not specifically identify and track emergency work performed by either Corps or industry vessels.

In a June 2000 report to the Congress, the Corps stated that the placement of the *Wheeler* in ready reserve had been a success and recommended that the vessel remain in ready reserve. However, the report contained a number of analytical and evidentiary shortcomings, and, when asked, the Corps could not provide any supporting documentation for its recommendation. In addition, the report also proposed that the *McFarland* be placed in ready reserve, but the Corps did not conduct an analysis to support this proposal. Furthermore, according to the Corps, the *McFarland* will require at least a $25 million capital investment to ensure its safety, operational reliability, and effectiveness for future service. It is questionable whether such an investment in a vessel that would be placed in ready reserve and receive only minimal use is in the best interest of the government.

We are making recommendations to the Secretary of the Army regarding the need to comprehensively analyze the costs and benefits of existing and proposed restrictions on the use of the Corps' hopper dredge fleet and to update the information and methodology that the Corps uses for its hopper dredging cost estimates. In commenting on a draft of this report, the Department of the Army agreed with all the recommendations and provided time frames for implementing each of them. The Dredging Contractors of America generally agreed with our recommendations, but strongly disagreed that the benefits of the restrictions are unproven.

BACKGROUND

Since 1824 the Corps has been responsible for constructing and maintaining a safe, reliable, and economically efficient navigation system. Today, this system is comprised of more than 12,000 miles of inland waterways, 300 large commercial harbors, and 600 small harbors. From fiscal years 1998 through 2002, the Corps has removed an average of about 265 million cubic yards of material each year from the navigable waters of the United States, at an average annual cost of about $856 million (in constant 2002 dollars). Private industry performs most of the overall dredging, except for the work done by hopper dredges, in which both the Corps and industry perform a significant amount of the work. Of the $856 million spent annually on overall dredging, about $197 million is spent on all hopper dredging (both maintenance and new construction), with industry vessels accounting for about $148 million annually and Corps vessels accounting for about $49 million.

Each of the Corps' hopper dredges typically operates in a specific geographic area. The *Wheeler*, a large-class dredge,[1] usually operates in the Gulf of Mexico. The *McFarland*, a medium-class dredge, usually operates in the Atlantic and Gulf of Mexico. The *Essayons*, a large-class dredge, and the *Yaquina*, a small-class dredge, typically work along the Pacific coast.

Legislation enacted in the 1990s sought to further increase the role of industry in hopper dredging by placing operational restrictions on the Corps' hopper dredges. Specifically, the Energy and Water Development Appropriations Act for fiscal year 1993 and subsequent appropriations acts required the Corps to offer for competitive bidding 7.5 million cubic yards of hopper dredging work previously performed by the federal fleet. Since fiscal year 1993, the Corps has addressed this requirement by reducing the use of each of its four dredges from about 230 workdays per year to about 180 workdays per year. The Water Resources Development Act for fiscal year 1996 required the Corps to initiate a program to increase the use of private hopper dredges principally by taking the *Wheeler* out of active status and placing it into ready reserve. The Corps implemented this requirement by allowing the *Wheeler* to work 55 days a year plus emergencies (which includes urgent and time-sensitive dredging needs). The 1996 act did not alter the Corps' duty to implement the dredging program in the manner most economical and advantageous to the United States, and it restricted the Corps' authority to reduce the workload of other federal hopper dredges. The conference report that accompanied the act directed the Corps to periodically evaluate the effects of the ready reserve program on private industry and on the Corps' hopper dredge costs, responsiveness, and capacity. The Energy and Water Appropriations Act for fiscal year 2002 placed another restriction on the use of the Corps' dredge *McFarland*, limiting it to emergency work and its historical scheduled maintenance in the Delaware River (about 85 workdays per year). Taken together, these restrictions have increased private industry's share of the hopper dredging workload.

In theory, restrictions on the use of the Corps' hopper dredges could generate efficiency and cost-savings benefits to both government and industry. For example, restricting the Corps' hopper dredges to fewer scheduled workdays could make them more available to respond to emergency dredging needs. In addition, the increase in demand for dredging by private industry could lead to improvements in dredging efficiency. If achieved, firms might be able to dredge the same amount of material at a lower cost or more material at the same cost. Furthermore, if more work were provided to the private hopper dredging industry, competition could increase if the existing dredging firms expanded their fleets or more firms entered the market.[2] Consequently, the prices that the government pays to contractors could fall. However, economic principles also suggest that if an industry is given more work without increasing capacity or the number of competing firms, prices could rise because the demand for its services has increased.

CORPS HAS TRANSFERRED MOST OF ITS HOPPER DREDGING TO PRIVATE INDUSTRY, BUT STILL NEEDS TO RETAIN A HOPPER DREDGE FLEET

The Corps' and private industry's respective roles in the hopper dredging market have changed since legislation enacted in 1978 prompted a movement toward privatization of hopper dredging in the United States. Since that time, the Corps has gradually reduced its hopper dredging fleet from 14 to 4 vessels, while a private hopper dredging industry of five firms and 16 vessels has emerged. Corps officials and representatives from the dredging industry, selected ports, and the maritime industry generally agreed that the Corps needs to retain at least a small hopper dredge fleet to (1) provide additional dredging capacity during peak demand years, (2) meet the emergency and national defense needs identified in the 1978 legislation, and (3) provide an alternative work option at times when the industry offers unreasonable bids or no bids at all. To determine the reasonableness of private contractor bids, the Corps develops a government cost estimate for its hopper dredging solicitations. If the low bid is no more than 25 percent above the government cost estimate, the Corps awards the contract. If all bids exceed the government cost estimate by more than 25 percent, the Corps may pursue a number of options, including performing the work itself. The practical value of this protection against high bids, however, has been limited by the Corps' use of some outdated contractor cost information and its continued use of an expired policy to calculate transit costs.

Corps and Industry Roles in Hopper Dredging Have Shifted

Before 1978, the Corps performed all of the nation's hopper dredge work. In 1978, the Congress passed legislation to encourage private industry participation in all types of dredging and required the Corps to reduce the fleet of federal vessels to the minimum necessary for national defense and emergency purposes, as industry demonstrated its capability to perform the work. According to the Senate committee report associated with the 1978 legislation, one of the law's main purposes was to provide incentives for private industry to construct new hopper dredges. Between 1978 and 1983, as a private hopper dredging industry began to emerge, the Corps reduced its hopper dredge fleet from 14 to its current 4 vessels. By the late 1980s, the Corps stopped assigning its hopper dredges to new construction projects (primarily channel deepening), leaving this work entirely to private industry. Both Corps and private industry hopper dredges continue to perform maintenance work on existing channels.

From fiscal years 1998 through 2002, the Corps' dredges performed about 28 percent of the nation's hopper dredging maintenance work, annually dredging about 16 million cubic yards of material at a cost of about $49 million (in constant 2002 dollars). During the same period, industry dredges performed about 72 percent of the nation's hopper dredging maintenance work, dredging about 40 million cubic yards of material annually, at a cost of about $93 million.[3] As a result of the 1978 legislation, seven firms emerged to compete for the Corps' hopper dredging contracts. Consolidation and firm buy-outs in the 1990s have left five firms in today's market. (Appendix II contains a more detailed description of the U.S. hopper dredge fleet.)

Corps' Hopper Dredge Fleet Addresses Several Needs

Corps officials and representatives from the dredging industry, selected ports, and the maritime industry generally agreed that the Corps' hopper dredge fleet currently (1) provides additional dredging capacity during peak demand years, (2) meets emergency dredging and national defense needs identified in the 1978 legislation, and (3) provides an alternative work option when industry provides no bids or when its bids exceed the government cost estimate by more than 25 percent. In addition, representatives of selected ports and the maritime industry generally supported the Corps' retention and operation of a federal hopper dredge fleet to ensure that dredging needs are met in a timely manner.

One of the reasons for the Corps to maintain a hopper dredge fleet is that changes in annual weather patterns, such as *El Niño*, and severe weather events, such as hurricanes and floods, can create a wide disparity in the demand for hopper dredging services from year to year. During fiscal year 1997 the Corps and private industry used their hopper dredges for maintenance work to remove almost 77 million cubic yards nationwide. In contrast, during fiscal year 2000 they removed about 50 million cubic yards. (See figure 2.) Hopper dredging needs at the mouth of the Mississippi River are particularly variable from year to year, with annual dredging requirements ranging from 2 million to 50 million cubic yards. Representatives from private dredging firms maintain that industry is not likely to build the additional capacity needed to meet demand in peak years. Corps officials and representatives from the dredging industry, selected ports, and the maritime industry generally agreed that the federal government should provide the additional dredging capacity required to meet the needs of peak demand years.

The Corps' hopper dredges are also needed to respond to emergency dredging assignments. For example, according to a Corps official, it was necessary for the Corps to send the *Essayons* to finish work on a project in Alaska that was critical to complete before the winter season and freezing conditions set in. In addition, Corps vessels have been used during instances where industry has submitted no bids in response to solicitations. For example, when rains in the Mississippi River Basin caused a build-up of material in navigation channels, the Corps issued a solicitation, but no bids were received because industry vessels were unavailable. Consequently, the *Wheeler* was used to perform the work. In such situations, the Corps' fleet acts as insurance to meet dredging needs, ensuring that shipping patterns are not adversely affected.

The existence of the Corps' fleet theoretically offers a measure of protection against inordinately high bids from private contractors. While the private dredging market consists of 16 dredges owned by five firms, not all dredges compete for any given solicitation because (1) some, if not most, hopper dredges are committed to other jobs; (2) hopper dredges may be in the shipyard; (3) differences in hopper dredge size and capability mean that not all hopper dredges are ideally suited to perform the work for a particular job; and

(4) hopper dredges cannot quickly move from one dredging region to another.[4] For example, large hopper dredges may have difficulty maneuvering in small inlet harbors, and small hopper dredges may be inefficient at performing large projects with distant disposal sites. Thus, the Corps' hopper dredge fleet provides an alternative dredging capability that can be brought to bear when private dredges are not available or when private industry bids are deemed too high.

Corps Bases Its Cost Estimate on Outdated Information

The Corps' government cost estimate for hopper dredging work is pivotal in determining the reasonableness of private contractor bids. The Corps is required to determine a fair and reasonable estimate of the costs for a well-equipped contractor to perform the work. By law, the Corps may not award a dredging contract if the price exceeds 25 percent of the government estimate. In such cases, the Corps has several options. It can (1) cancel the solicitation, (2) readvertise the solicitation, (3) consider challenges to the accuracy of the Corps' cost estimate by bidders, (4) convert the solicitation into a negotiated procurement, or (5) use one of its own dredges to do the work.

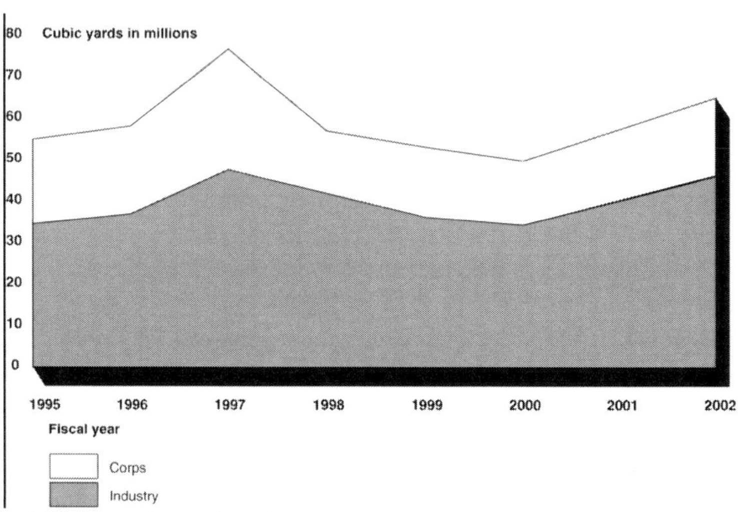

Source: U.S. Army Corps of Engineers.
Note: GAO analysis of Corps' Navigation Data Center data.

Figure 2. Maintenance Hopper Dredging Work, Fiscal Years 1995 through 2002.

The accuracy of the Corps' cost estimate depends on having access to upto-date cost information. Although the Corps adjusts contractor cost data annually to rcflcct currcnt pricing lcvcls, this stcp docs not account for fundamental changes, such as an industry vessel reaching the end of its depreciable life or industry acquisition of new vessels. The Corps has not obtained comprehensive industrywide contractor cost information since 1988. Since then, contractors have provided the Corps updated cost information to support specific costs included in the Corps' cost estimates that they believe to be outdated, but they are not required to provide updated information for all costs. As a result, the Corps only has updated cost information that contractors provide. In our discussions with Corps officials, they acknowledged the need to initiate an effort to obtain and verify current cost data for industry vessels.

In addition, the Corps continues to follow an expired policy when calculating contractor transit costs to the dredge site, further calling into question the accuracy of the government cost estimates. The Corps' Engineering Regulation 1110-2-1300, which called on the Corps to calculate industry transit costs to the dredge site based on the location of the second-closest industry dredge, expired in 1994.[5] However, the Corps continues to use this method when calculating transit costs for at least some of its solicitations. For example, Corps officials followed the expired policy when demonstrating to us how they calculated the transit costs for a solicitation in Washington State.[6] In this case, the secondclosest industry dredge was located in the Gulf of Mexico, and the estimated transit costs amounted to about $480,000 because the vessel would have had to travel thousands of miles and go through the Panama Canal. However, the private contractor's dredge that performed the work was located fewer than 500 miles from the dredge site, for which the transit costs were estimated to be about $100,000. After bringing this issue to the Corps' attention, the Corps told us that it plans to reexamine its transit cost policies.

RESTRICTIONS ON THE CORPS' HOPPER DREDGE FLEET HAVE IMPOSED COSTS, BUT BENEFITS ARE UNPROVEN

Restrictions on the Corps' hopper dredge fleet, which began in fiscal year 1993, have imposed costs on the Corps' dredging program, but have thus far not resulted in proven benefits. Most of the costs of the Corps' hopper dredges are incurred regardless of how frequently the dredges are used. A possible benefit of the restrictions is that they could eventually encourage more firms to enter the market or existing firms to add capacity, which, in turn, may promote competition, improve dredging efficiency, and thus reduce prices. Although there has been an increase in the number of private industry hopper dredges since the restrictions were first imposed, the number of private firms in the hopper dredging market has decreased. In addition, during the same time period, the number of contractor bids per Corps solicitation has decreased, while the number of winning bids exceeding the Corps' cost estimate has increased. Restrictions on the Corps' vessels could also potentially enhance the Corps' responsiveness to emergency dredging needs. However, the Corps is unable to evaluate whether emergency dredging needs have been met more or less efficiently since the restrictions went into effect because it does not specifically identify and track emergency work performed by either Corps or industry vessels.

Corps Incurs Costs by Keeping the *Wheeler* in Ready Reserve

The Corps incurs many of the costs for maintaining and operating its hopper dredges regardless of how much the vessels are used. Thus, as shown in table 1, when the *Wheeler* was placed in ready reserve and restricted to 55 workdays plus emergencies, the average number of days it worked per year and its productivity (measured by cubic yardage dredged) declined by about 56 percent, while its costs declined by only 20 percent. Crew size declined by about 21 percent, but payroll costs declined by just 2 percent because dredging needs required the Corps to pay the smaller crew overtime to finish the work. In addition, fuel costs did not drop in proportion to use and productivity because the vessel's engines were utilized for shipboard services (e.g., electricity) while it remained at the dock—a necessary procedure for maintaining minimal vessel readiness. Other costs unrelated to crew or fuel represent the plant or capital costs of a dredge, many of which the Corps incurs regardless of how much a dredge is used.

The Corps refers to the difference between a vessel's total costs and the value of the dredging services it provides (the net cost) as a "subsidy." The Corps estimates the annual subsidy to maintain the *Wheeler* idle in ready reserve at about $8.4 million.[7] This subsidy is a direct cost of ready reserve. In addition to the subsidy, the Corps must pay contractors to do the work the *Wheeler* no longer performs. The difference between the vessel's traditional workload and its current workload is approximately 6.6 million cubic yards.

Depending on whether private industry hopper dredges are able to perform this work in aggregate at a lower or higher cost than if the *Wheeler* performed the work, the total cost to government of the *Wheeler* in ready reserve status could be either lower or higher than the Corps' estimated subsidy.

In addition to the *Wheeler's* subsidy, restrictions have led to inefficient operations for the other Corps hopper dredges, resulting in additional costs for the Corps. According to Corps officials, September is the ideal time to dredge in the Pacific Northwest, because dredging conditions generally deteriorate in October. The officials mentioned that, at times, the *Essayons* and the *Yaquina* have reached their fiscal year operating limits and returned to port in September, before the projects they were working on were complete. The dredges were sent back to complete the project after the new fiscal year began in October, even though weather conditions may have made dredging conditions less than optimal, and the Corps incurred additional transit costs. According to some Corps officials, the annual operating limit cannot be extended. For example, the *Essayons* stopped dredging the mouth of the Columbia River and returned to port at the end of fiscal year 2001 when it reached its operating limit. The vessel returned to finish the work at the start of the new fiscal year, but adverse weather conditions prevented it from fully dredging the river. As a result, some projects may be postponed until the following fiscal year, reprioritized, or canceled altogether.

Table 1. Summary of Operations and Cost Data of Corps' Dredge *Wheeler*

Component	Before reserve[a]	After reserve[b]	Percentage change
Average days worked	183	83	-55%
Average cubic yards	11,847,040	5,245,606	-56%
Crew size	54	42	-21%
Payroll costs Average cost[c]	$3,635,146 $17,136,028	$3,557,938 $13,631,862	-2%-20%
Fuel costs	$1,206,578	$832,452	-31%
Other costs[d]	$12,294,304	$9,241,472	-25%

Source: U.S. Army Corps of Engineers.

Note: GAO analysis of data obtained from the Corps' *Annual Form 27 Report of Operations—Hopper Dredges*, fiscal years 1994 through 2001.

[a] Fiscal years 1994 through 1997 represent the time period before ready reserve.

[b] Fiscal years 1998 through 2001 represent the time period after ready reserve.

[c] In constant 2001 dollars.

[d] These costs include, among other things, depreciation and repairs.

Benefits of Restrictions Are Unproven

A potential benefit of the restrictions on the Corps' hopper dredge fleet is that an increase in demand for industry's dredging services could encourage existing firms to make capital investments (e.g., build new dredges or improve existing dredges) or encourage more firms to enter the dredging market. Dredging industry representatives told us that the restrictions have already led to an increase in the number of industry vessels and, as evidence, pointed to the addition of two new dredges, the *Liberty Island*, a large-class dredge introduced in 2002, and the *Bayport*, a medium-class dredge introduced in 1999, as well as the return of the *Stuyvesant*, a large-class dredge, to the U.S. hopper dredging market. Moreover, they added that since the restrictions, the private hopper dredging industry has also made improvements and enhancements to its existing fleet—specifically the *Columbia*—thus improving the efficiency of its dredging operations and increasing the capacity of its vessels. However, the representatives also told us that the restrictions are only one of several factors the private hopper dredging industry considers when deciding to acquire or build an additional dredge. In addition, firms must invest in equipment to remain competitive in any industry. As a result, it is unclear to what extent the restrictions on the Corps' vessels were a factor in industry's investment decisions to increase its fleet size and add dredging capacity.

While the private hopper dredging industry has recently placed two new dredges on line, it has sold the small-class dredge *Mermentau* and placed another small-class dredge, the *Northerly Island*, up for sale. In addition, during the last decade the private hopper dredging industry has decreased from seven firms to five firms. Specifically, since 1993, two firms exited the market, one firm entered the market, and two firms merged. The consolidation in the industry does not necessarily mean that competition has been reduced because the new industry structure could have resulted in enhanced capacity, flexibility, and efficiency for the remaining firms.

However, it is uncertain whether the private hopper dredging industry is more or less competitive now than it was prior to the restrictions. Historical data reveal that, in general, as shown in figure 3, in years when more material is available to private industry, industry submits fewer bids per Corps solicitation. For example, during fiscal year 1991, when the Corps estimated that 31.3 million cubic yards of maintenance material would be contracted out to industry, the average number of bids per solicitation was 3.2. In contrast, during fiscal year 1998, when the Corps estimated that 53.7 million cubic yards of maintenance material would be contracted out to industry, industry submitted an average of about 2 bids per solicitation.

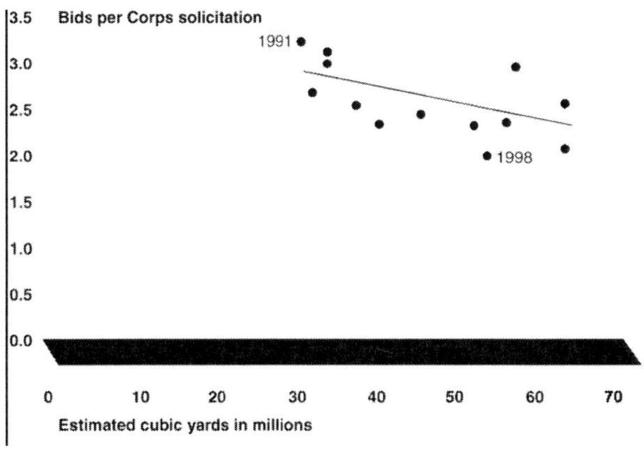

Source: U.S. Army Corps of Engineers.
Notes: GAO analysis of the Corps' Dredging Information System data.
Each point represents a fiscal year.
The inverse, linear relationship is statistically significant at the 95 percent level.

Figure 3. Estimated Volume of Material Dredged by Industry and Average Number of Industry Bids per Corps Solicitation, Fiscal Years 1990 through 200.

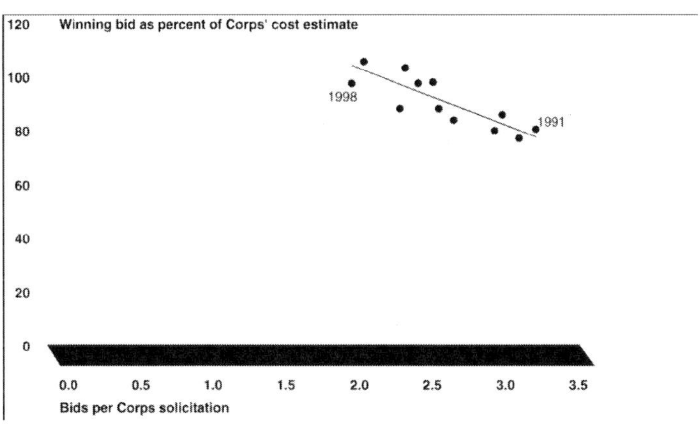

Source: U.S. Army Corps of Engineers.
Notes: GAO analysis of the Corps' Dredging Information System data.
Each point represents a fiscal year.
The linear relationship is statistically significant at the 99 percent level.

Figure 4. Annual Average Number of Industry Bids per Corps Solicitation and Winning Bid as a Percentage of the Corps' Cost Estimate, Fiscal Years 1990 through 2002.

Likewise, as shown in figure 4, in years when there were fewer industry bids per Corps solicitation, the average winning industry bid, as a percentage of the Corps' cost estimate, was higher.[8] For example, during fiscal year 1991, when the average number of bids per solicitation was 3.2, the average winning bid was 79 percent of the Corps' estimate. In contrast, during fiscal year 1998, when the average number of bids per solicitation was 2, the average winning bid was 97 percent of the Corps' estimate.

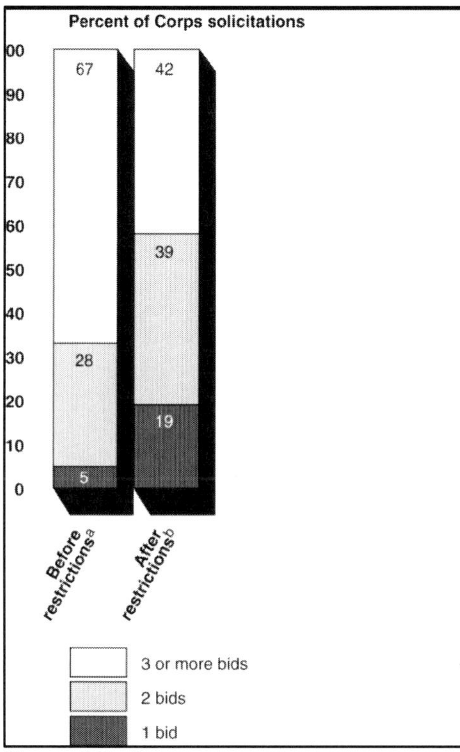

Source: U.S. Army Corps of Engineers.

Note: GAO analysis of the Corps' Dredging Information System data.

[a] Fiscal years 1990 through 1992 represent the time period before the restrictions were implemented.

[b] Fiscal years 1993 through 2002 represent the time period after the restrictions were implemented.

Figure 5. Comparison of Number of Industry Bids per Corps Solicitation Before and After Restrictions.

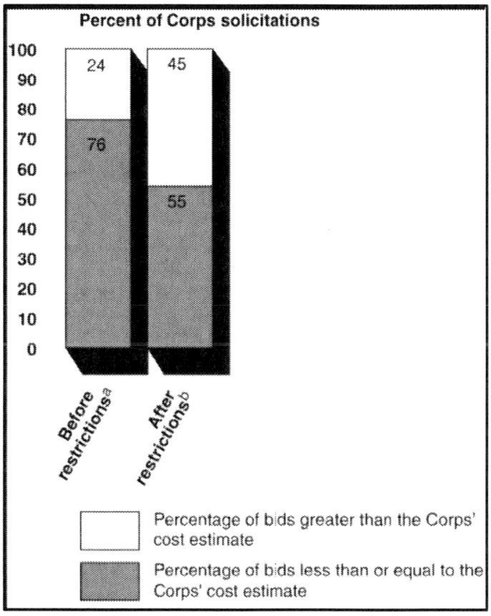

Source: U.S. Army Corps of Engineers.

Note: GAO analysis of the Corps' Dredging Information System data.

[a] Fiscal years 1990 through 1992 represent the time period before the restrictions were implemented.

[b] Fiscal years 1993 through 2002 represent the time period after the restrictions were implemented.

Figure 6. Comparison of Winning Industry Bids and Corps' Cost Estimates Before and After Restrictions.

In general, when there are fewer industry bids per solicitation, the winning industry bid relative to the Corps' cost estimate increases. In fiscal years 1990 through 2002, more than half of the solicitations for hopper dredging maintenance work received just one or two bids from private contractors. During these years, when only one contractor bid on a solicitation, the bid exceeded the government estimate 87 percent of the time. In contrast, when there were three or more bids on a solicitation, the winning bid exceeded the government estimate only 22 percent of the time. After the Corps' hopper dredge fleet was effectively restricted to 180 workdays (fiscal years 1993 through 2002), the number of industry bids per solicitation declined from about 3 to roughly 2.4. Specifically, as shown in figure 5, when there were no limits on the use of the Corps' hopper dredges (fiscal years 1990 through 1992), only 5 percent of solicitations received one bid. After limits were placed on the Corps' hopper dredges (fiscal years 1993 through 2002), 19 percent of solicitations had only one bid.[9] Moreover, before the restrictions, 67 percent of the solicitations had three or more bids, whereas, after the restrictions, only 42 percent had three or more bids. These changes might have been expected because, after the restrictions, industry's share of hopper dredging work increased while the number of hopper dredging firms decreased from seven to five.

Furthermore, in the time period following the imposition of the 180-day restriction, the frequency with which the winning industry bid exceeded the Corps' cost estimate has increased. For example, as shown in figure 6, prior to the restrictions, the winning bid exceeded the Corps' cost estimate 24 percent of the time. After the restrictions were imposed, the winning bid exceeded the Corps' estimate 45 percent of the time.[10] This finding is consistent with economic principles; that is, all else equal, an increase in demand for dredging by private industry with fixed supply would result in higher prices.

It should be noted that the extent to which the restrictions contributed to the decrease in the number of industry bids per Corps solicitation and the increase in the winning industry bid relative to the Corps' cost estimate is unknown. Other factors could have also contributed to these changes. For example, an increase in the demand for hopper dredging services for new construction projects or beach nourishment could lead to a decrease in the number of bids received for maintenance projects. Similarly, the introduction of environmental restrictions on when hopper dredging can take place could contribute to an increase in the winning industry bid relative to the Corps' cost estimate. Nevertheless, the decrease in the number of bids per solicitation and the increase in bids exceeding the Corps' cost estimates raises questions about the effects the restrictions may have had on competition and prices, demonstrating the need for a comprehensive analysis of the effects of the restrictions on competition, efficiency, and prices.

Another potential benefit of restrictions on the use of the Corps' vessels is enhanced responsiveness to emergencies. However, there is disagreement within the Corps on this issue. One Corps official believes that a dredge in ready reserve is better able to handle emergencies than if it were working 180 days because it is in a "standby" status at the dock, ready to respond. In contrast, others in the Corps believe that a dredge can respond just as well or better to an emergency while working a full schedule because the dredge can temporarily halt the project it is working on, respond to the emergency, and then return to its scheduled work.[11] During our discussions with representatives from selected ports and the maritime industry, we did not learn of any instances of problems in the Corps' responsiveness to emergencies prior to restrictions or instances of improved response time since the restrictions went into effect.

A major reason that the Corps is unable to evaluate whether emergency dredging needs have been met more or less efficiently since the restrictions went into effect is that its dredging database—the Dredging Information System—does not specifically identify and track emergency work performed either by Corps or industry vessels. Consequently, the Corps cannot readily determine how many days have been needed for each of its vessels to respond to emergencies. In addition, the Corps does not know whether it is paying contractors more or less for performing the emergency dredging projects compared to the costs it pays for routinely scheduled maintenance work. Such information would be a valuable tool for determining how emergency dredging needs can be met in a manner that is the most economical and advantageous to the government—that is, when and under what circumstances to contract with the private hopper dredging industry for these emergencies or when to use Corps vessels. In discussing this issue, Corps officials agreed that obtaining information on emergencies is important for managing their hopper dredging program and told us they have initiated efforts to collect such data to incorporate into their dredging database.

Corps Has Not Justified the Existing and Proposed Restrictions on Its Hopper Dredge Fleet

In a June 2000 report to the Congress, the Corps stated that the placement of the *Wheeler* in ready reserve had been a success and recommended that the vessel remain in ready reserve. However, the report contained a number of analytical and evidentiary shortcomings, and, when asked, the Corps could not provide any supporting documentation for its recommendation. In addition, the report also proposed that the *McFarland* be placed in ready reserve, but the Corps did not conduct an analysis to support this proposal. The costs to place the *McFarland* in ready reserve are likely to be similar to the costs incurred by placing the *Wheeler* in ready reserve. Because the *McFarland's* workload would be reduced from 180 days to 55 days plus emergencies, the Corps would incur annual costs of about $8 million when the vessel is idle—largely because much of a vessel's costs are incurred regardless of its level of use. Furthermore, according to the Corps, the *McFarland* will require at least a $25 million capital investment to ensure its safety, operational reliability, and effectiveness for future service. It is questionable whether such an investment in a vessel that would be placed in ready reserve and receive only minimal use is in the best interest of the government.

Corps Has Not Comprehensively Analyzed the Costs and Benefits of Restrictions

The Water Resources Development Act for 1996 required the Corps to determine whether (1) the *Wheeler* should be returned to active status or continue in ready reserve status or (2) another federal hopper dredge should be placed in ready reserve status, and issue a report to the Congress on its findings. The Corps issued the required report in June 2000,[12] recommending that the *Wheeler* remain in reserve and proposing that an additional dredge, the *McFarland*, also be placed in reserve. However, when asked, the Corps official who authored the report told us that he did not have any supporting documentation for the report. In addition, the report had a number of evidentiary and analytical shortcomings. For example, the evidence presented in the report showed that the price the government paid to the industry for hopper dredging was higher in the 2 years after the *Wheeler* was put in ready reserve than it was the year before. This raises questions about the validity of the recommendation contained in the report.

Furthermore, the report did not contain a comprehensive analysis. A comprehensive economic analysis of a government program or policy would identify all the resulting costs and benefits, and, where possible, quantify these measures. Both the quantitative and qualitative costs and benefits would need to be compared and evaluated to determine the success or failure of a program and to potentially be used as a basis for future policy decisions. With regard to the restrictions on the Corps' hopper dredges, a comprehensive economic analysis might contain, among other things, all costs associated with the nonuse of the vessel and the potential benefits that might result due to efficiency gains, increased competition, and lower prices. The analysis might also examine whether ports, harbors, and access channels were maintained more or less effectively, or whether emergency dredging needs were met in a more or less timely and cost-effective manner following implementation of the restrictions.

Corps Has Not Demonstrated that Further Restrictions on the Use of the Corps' Dredge *McFarland* Are Warranted

The Corps has not demonstrated that placing an additional hopper dredge in ready reserve, specifically the *McFarland*, would be beneficial to the United States. In its June 2000 report to the Congress on the ready reserve status of the dredge *Wheeler*, the Corps proposed that the *McFarland* be the next dredge placed in reserve. However, the Corps did not offer any analysis on the potential costs of placing an additional Corps hopper dredge in reserve or the benefits of such an action. Moreover, to be available for future use, the 35-year-old *McFarland* requires at least a $25 million capital investment to ensure its safety, operational reliability, and effectiveness. The repairs include asbestos removal; repairs to the hull; engine replacement; and upgrades of equipment, machinery, and other shipboard systems. It is questionable whether spending $25 million to rehabilitate the *McFarland* and then placing it in ready reserve is prudent.

Furthermore, if the *McFarland* were placed in ready reserve, the Corps would incur annual costs similar to the subsidy that is already incurred for the *Wheeler*. Because the *Wheeler's* costs do not vary proportionally to its use, the costs to operate the vessel 55 days a year plus emergencies in ready reserve is only marginally less than if it were to operate 180 days a year. The Corps estimates that if the *McFarland* were placed in ready reserve, it would require an annual subsidy of about $8 million to remain idle. The Corps would also need to contract out the work the *McFarland* would no longer be doing—approximately 2 to 3 million cubic yards per year. Depending on whether private industry hopper dredges are able to perform this work in aggregate at a lower or higher cost than if the *McFarland* performed the work, the total cost to government of the placing the *McFarland* in reserve could be either lower or higher than the estimated annual subsidy. Finally, placing the *McFarland* in ready reserve could increase competition if such restrictions spurred an increase in investment in private hopper dredges. However, it is questionable whether placing the *McFarland* in ready reserve would provide enough incentive for industry to make additional investments.

CONCLUSION

Hopper dredges play a critical role in keeping the nation's ports open for both domestic and international trade. This function has been and will likely continue to be carried out through a mix of private industry and government-owned dredges. At issue is how to use this mix of dredges in a manner that maintains the viability of the private fleet while minimizing the costs to government. The Corps has proposed to the Congress that additional restrictions on the use of its hopper dredges are warranted, but it cannot provide any analytical evidence to support its position. The limited evidence that does exist indicates that these restrictions have imposed costs on the government, while the benefits are largely unproven. Unless and until the Corps gathers the data, comprehensively analyzes the costs and benefits of restrictions on the use of its hopper dredges, and takes the steps to update its cost estimates, there is no assurance that the nation's hopper dredging needs are being met in a manner that is the most economical and advantageous to the government.

Recommendations for Executive Action

In an effort to discern the most economical and advantageous manner in which to operate its hopper dredge fleet, and because the Corps has been unable to support, through analysis and documentation, the costs and benefits of placing its hopper dredges in ready reserve, we recommend that the Secretary of the Army direct the Corps of Engineers to

- obtain and analyze the baseline data needed to determine the appropriate use of the Corps' hopper dredge fleet including, among other things, data on the frequency, type, and cost of emergency work performed by the Corps and the private hopper dredging industry; contract type; and solicitations that receive no bids or where all the bids received exceeded the Corps' estimate by more than 25 percent;
- prepare a comprehensive analysis of the costs and benefits of existing and proposed restrictions on the use of the Corps' hopper dredge fleet—including limiting the Corps' dredges to 180 days of work per year, placing the *Wheeler* into ready reserve, limiting the *McFarland* to its historic work in the Delaware River, and placing the *McFarland* into ready reserve status; and
- assess the data and procedures used to perform the government cost estimate when contracting dredging work to the private hopper dredging industry, including, among other things, (1) updating the

cost information for private industry hopper dredges and (2) examining the policies related to calculating transit costs.

AGENCY COMMENTS

We provided a draft of this report to the Acting Assistant Secretary of the Army and the Dredging Contractors of America for review and comment.

In a letter dated March 21, 2003, the Department of the Army (Army) provided comments on a draft of this report. The Army agreed with our recommendations and provided time frames for implementing each of them. It also provided additional comments suggesting clarification and elaboration on a number of issues discussed in our report.

In a letter dated March 3, 2003, the Dredging Contractors of America (DCA) provided detailed comments on a draft of this report. DCA generally agreed with our recommendations. However, it believed strongly that reducing the scheduled use of the Corps' hopper dredges has resulted in proven benefits. We continue to believe that the relationship between the restrictions and benefits to the government are unproven because (1) the Corps incurs costs related to the underutilization of its dredges, and (2) since the restrictions were first imposed, the Corps has received fewer industry bids per solicitation, and the percentage of winning industry bids that exceed the Corps' cost estimates has increased.

We conducted our review between January 2002 and February 2003 in accordance with generally accepted government auditing standards. A detailed discussion of our scope and methodology is presented in appendix I.

Barry T. Hill
Director, Natural Resources and Environment

APPENDIX I: SCOPE AND METHODOLOGY

To assess the changing roles of the Corps and industry in hopper dredging and the characteristics of the hopper dredging industry, we obtained Corps' studies and data from the Corps' Navigation Data Center that provided information on the hopper dredging requirements of the United States, including the quantity of material dredged annually by the Corps and the private hopper dredging industry, and their associated costs. We also reviewed the laws that define these roles. In addition, we interviewed Corps officials; representatives from the five hopper dredging firms (B+B Dredging Co., Inc., Bean Stuyvesant LLC, Great Lakes Dredge & Dock Company, Manson Construction Co., and Weeks Marine, Inc.); the maritime industry (the Delaware River Port Authority, Maritime Exchange for the Delaware River and Bay, Navios Ship Agencies, Inc., and the Steamship Association of Louisiana); dredging and port associations (Dredging Contractors of America, Pacific Northwest Waterways Association, and American Association of Port Authorities); and selected ports (Portland, Seattle, New York/New Jersey, New Orleans, and Wilmington). To obtain a better understanding of hopper dredging from the perspective of the private hopper dredging industry, we visited and toured a medium-class industry hopper dredge working in the Chesapeake and Delaware Canal and interviewed its crew. Moreover, we reviewed the Corps' cost estimating policies.

To determine the intent and effects of the restrictions placed on the use of the Corps' hopper dredge fleet, we analyzed the laws governing the use of the Corps' hopper dredges. We also reviewed studies conducted by the Corps and the Pacific Northwest Waterways Association. For qualitative information, we obtained documents and interviewed Corps officials from headquarters and district and division offices, including Jacksonville, New Orleans, Philadelphia, Portland, Walla Walla, and the North Atlantic Division, as well as representatives from the private hopper dredging firms, selected ports, dredging and port associations, and the maritime industry. For quantitative information, we performed descriptive statistical analyses using data on the winning contractor bids, estimated industry dredging volumes, and the Corps' cost estimate available from the Corps' Dredging Information System database.

To evaluate whether further restrictions on the Corps' hopper dredge fleet, including placing the Corps' dredge *McFarland* in ready reserve, are justified, we reviewed studies and analyses performed by the Corps to support its proposal to place the *McFarland* in ready reserve. We also interviewed officials from the Corps and representatives from the private hopper dredging industry, selected ports, and the maritime industry to gain their views on the possible effects on competition and emergency response if the current restrictions on the Corps' hopper dredges, particularly the *McFarland*, were modified. To determine the costs associated with repairing the *McFarland*, we obtained and analyzed cost estimates for the repairs prepared by the Corps' Philadelphia district office and discussed the estimates with Corps district and headquarters officials. We also visited and toured the *McFarland* when it was working in the Delaware River and interviewed the *McFarland's* crew and Corps officials from the Philadelphia district and the North Atlantic Division offices.

We conducted our review between January 2002 and February 2003 in accordance with generally accepted government auditing standards.

APPENDIX II: THE U.S. HOPPER DREDGE FLEET

There are currently 20 hopper dredges operating in the United States. (See table 2.) Of the 20 dredges, 4 are small-class hopper dredges, 10 are medium-class hopper dredges, and 6 are large-class hopper dredges. Of the 16 private hopper dredges, Great Lakes Dredge & Dock Company owns 7, Manson Construction Co. owns 3, and the remaining firms (B+B Dredging Co., Inc., Bean Stuyvesant LLC, and Weeks Marine, Inc.) each own 2.

Table 2. Corps and Private Industry Hopper Dredge Fleets

Size	Owner	Vessel	Capacity (in cubic yards)	Year built
Large-	Great Lakes Dredge & Dock	*Liberty Island*	6,540	2002
class	Company	*Long Island*	16,000	1971
	Bean Stuyvesant LLC	*Stuyvesant*	11,200	1982
		Eagle I	6,600	1981
	Corps of Engineers	*Wheeler*	8,256	1982
		Essayons	6,000	1983
Medium-class	B+B Dredging Co., Inc.	*Columbia*	4,000	1986a
	Weeks Marine, Inc.	*B.E. Lindholm*	4,150	1985
		R.N. Weeks	4,000	1987
	Manson Construction Co.	*Bayport*	5,000	1999
		Newport	4,000	1983
	Great Lakes Dredge & Dock	*Dodge Island*	3,600	1980
	Company	*Manhattan Island*	3,600	1977

Table 2. (Continued)

Size	Owner	Vessel	Capacity (in cubic yards)	Year built
		Padre Island	3,600	1981
		Sugar Island	3,600	1979
	Corps of Engineers	*McFarland*	3,140	1967
Small-	Great Lakes Dredge & Dock	*Northerly Island*	2,160	1983
class	Company			
	Manson Construction Co.	*Westport*	1,800	1978
	B+B Dredging Co., Inc.	*Atchafalaya*	1,300	1980
	Corps of Engineers	*Yaquina*	1,020	1981

Source: U.S. Army Corps of Engineers.

[a] Although the vessel was originally built in 1944 to transport military equipment in World War II and later converted to a hopper dredge, according to Corps' data, 1986 is listed as the year the vessel began its service.

End Notes

[1] A hopper dredge's class is determined by its capacity—hoppers with up to 3,000 cubic yards of capacity are considered small, medium hoppers have capacity from 3,000 to 6,000 cubic yards, and large hoppers have a capacity of 6,000 cubic yards or more.

[2] Hopper dredging requires large capital outlays—a medium-class hopper dredge costs between $20 million and $40 million and normally takes 18 months to build—making it difficult for firms to enter the market quickly.

[3] A direct and valid comparison of the Corps' and private industry's costs to perform hopper dredge work is not possible due to various factors, which include, among other things, design features in the Corps' vessels in support of national defense missions, which add weight to the vessels and make them less efficient than industry vessels; limits to the number of days the Corps' vessels may operate—180 days or fewer, compared to about 250 days for industry; and differences between dredging projects—such as type of material dredged, type of work and corresponding risk level, and distance from the dredging operations to the disposal site.

[4] There are three main regions where hopper dredging takes place in the United States—the Atlantic, the Gulf of Mexico, and the Pacific. Dredges can move readily from the Atlantic to the Gulf of Mexico (which requires at least a week), but moving from the Atlantic to the Pacific requires several weeks and transit through the Panama Canal.

[5] In 1994 the Corps replaced the expired regulation with Corps' Engineering Regulation 1110-2-1302, which called on the Corps to base transit costs on a radius for a normal area of operations from the project site that includes a reasonable number of bidders.

[6] Transit costs have a greater impact on solicitations that take place in the Pacific Northwest, where the second-closest dredge may be more distant from the work site than for solicitations that take place in the Gulf of Mexico or the Atlantic.

[7] According to the Corps, the *Wheeler's* average annual operating cost during ready reserve was $12.5 million. While the vessel is credited for "earning" $4.125 million for its 55 days of work (at a daily rental rate of $75,000), a subsidy of $8.375 million per year is required to maintain the *Wheeler* idle in ready reserve.

[8] As previously discussed, we have identified concerns related to the Corps' cost estimate. However, these concerns were largely present both before and after the restrictions, thus we have no reason to believe that these concerns would materially affect the use of the cost estimate in the information presented.

[9] In the 5-year period following ready reserve of the *Wheeler* (fiscal years 1998 through 2002), there were roughly 2.5 industry bids per Corps solicitation, and 19 percent of the solicitations had only one bid, while 51 percent received three or more bids.

[10] In the 5-year period following ready reserve of the *Wheeler* (fiscal years 1998 through 2002), 42 percent of the winning bids exceeded the Corps' cost estimate, and 58 percent of the winning bids were less than the Corps' cost estimate.

[11] A vessel actively working will respond to an emergency with a full crew, whereas a vessel in reserve may be called to respond to an emergency during a period when it has a reduced crew and may be unable to assemble a full crew and respond to an emergency in a timely manner.

[12] *Report to Congress, Section 237, Hopper Dredges: Ready Reserve Status of the Hopper Dredge Wheeler.*

In: Hopper Dredging of Waterways
Editor: Eleanor G. Chance
ISBN: 978-1-61122-357-6

Chapter 3

REPORT TO CONGRESS: HOPPER DREDGES*

U.S. Army Corps of Engineers

EXECUTIVE SUMMARY

The Conference Report for the 2004 Energy and Water Appropriations Act directed the Corps of Engineers to report to the Appropriations Committees with a detailed plan of how it intends to rectify the issues raised in the March 2003 GAO report entitled "Effects of Restrictions on Corps' Hopper Dredges Should be Comprehensively Analyzed". The plan is to include how the Corps intends to establish a baseline for determining the appropriate use of the Corps hopper dredge fleet in the future. The conferees also requested that the Corps include a comprehensive analysis of the costs and benefits of the existing and proposed restrictions on the use of the fleet, expected the Corps to put in place measures by which better investment decisions regarding the fleet can be made.

The goal of this analysis was to identify the best business case that can ensure the ability to provide an acceptable level of service to the ports and users of the Nation's waterborne transportation system. The degree of change and configuration of the Corps minimum hopper dredge fleet was determined by resolving an acceptable level of risk both to the industry and to the ports, Corps partners and maritime users. In order to ensure that the industry dredges are consistently available to support the dredging requirements of the

* This is an edited, reformatted and augmented version of a report dated June 3, 2005.

Navigation Program, there must be a reasonable ability to realize a return on investment that will keep the vessels and equipment fully operational.

This report explains the navigation mission, the dredging environment, current problems, and issues concerning management of the four hopper dredges in the Corps minimum dredge fleet and the combined industry and Corps hopper dredge fleet. The data management issues raised in the GAO report are addressed and the improvements described. The analysis of costs and benefits of existing and proposed restrictions on the use of Corps hopper dredges is presented and the results are presented in terms of evaluations of twelve options.

The recommended option would schedule the Corps hopper dredge *Essayons* for 215 days, schedule the *Yaquina* for 178 days, keep the *Wheeler* in ready reserve, retire the *McFarland,* and continue with the initiatives from the Industry Corps Hopper Dredge Management Group. This Option would result in a $10.1 million net reduction in the total cost for hopper dredging, offer approximately 55 days additional work to industry, ensure that there is a viable reserve capability ready to respond to unforeseen requirements, and ensure that Federal projects requiring hopper dredging can be accomplished in a timely manner and at reasonable cost.

INTRODUCTION

The Fiscal Year 2004 Conference Report for the Energy and Water Appropriations Act made the following statement:

"During fiscal year 2002, the General Accounting Office [GAO] reviewed the benefits and effects of current and proposed restrictions on the Corps of Engineers' hopper dredge fleet. Congress faces significant future investments in the Corps hopper dredge fleet, as it is rapidly aging. The conferees believe that the investment decisions must take into consideration the subsequent use of the fleet. The final GAO report, released March 2003, reviewed the impacts of operational changes to the fleet since fiscal year 1993. GAO's findings made it clear to the conferees that additional costs have been imposed upon the Corps with the decreased use of the fleet, but that the benefits have not been realized. Additionally, the GAO found that the Corps' contracting process for hopper dredges was not effective. Most importantly, the GAO reported that the Corps did not have even a limited system to evaluate the costs and benefits of the varying operational levels of its hopper dredge fleet, nor did it have a means to make maintenance and repair decisions of the fleet taking operational use into consideration. The conferees remain concerned that since 2000, the Corps has provided to Congress, a report which has been found to have no analytical basis, thus calling into question the ready reserve policy.

Therefore, the conferees direct the Corps of Engineers to report to the Appropriations Committees within 6 months of enactment of this Act, with a detailed plan of how it intends to rectify the current situation. The plan is to include how the Corps intends to establish a baseline for determining the appropriate use of the Corps hopper dredge fleet in the future. Finally, the Corps shall include a comprehensive analysis of the costs and benefits of the existing and proposed restrictions on the use of the fleet. Overall, the conferees expect the Corps to put in place measures by which better investment decisions regarding the fleet can be made."

This report presents the U.S. Army Corps of Engineers response to Congress and response to the findings of the referenced March 2003 GAO report entitled "Effects of Restrictions on Corps' Hopper Dredges Should be Comprehensively Analyzed".

The analysis must take into account all of the variables affecting the hopper dredge requirements and capabilities, and must consider the costs of the equipment, the long-term costs, and the risks associated with each alternative. The analysis should also take into account that congressional direction and resultant Corps policy has been to move to greater use of the private sector. This analysis must begin with the premise that the results should improve the ability to perform the navigation mission in the most effective and efficient manner. This is not about Corps dredges or commercially owned and operated industry dredges, but about the combined fleet of hopper dredges comprised of Corps and industry. The goal of this analysis should be to identify the best business case that can ensure the ability to provide an acceptable level of service to the ports and users of the Nation's waterborne transportation system. The degree of change and configuration of the Corps minimum hopper dredge fleet will be determined by resolving an acceptable level of risk both to the industry and to the ports, Corps partners and maritime users. The majority of the hopper dredges in the combined fleet are owned by industry. It is important that the industry dredges are consistently available to support the dredging requirements of the Navigation Program. In order to ensure the availability of the industry, there must be a reasonable ability to realize a return on investment that will keep the vessels and equipment fully operational. The Corps must seek a way to schedule the industry work first, and stand ready to respond to peak workloads, emergency, and national defense needs. This determination is not founded in data or cost analysis alone, but must be considered in light of significant expectations for consistent maintenance of justified channel dimensions, and the fact that there is great variability in the dredging requirements in any given year. To expect industry to build a fleet that will ensure capability under the most extreme dredging demands is not an acceptable level of risk to impose upon the industry, nor is it a cost level that taxpayers would be willing to bear. This capability to deal with the peaks in workload is the primary objective of the Government's minimum dredge fleet. The challenge is to determine at what level of workload industry can not be relied on to perform at reasonable cost and in a timely manner, how likely and how frequently that workload can be expected to occur, and what the lowest possible number of Federal dredges is necessary to fill any gaps that industry can not handle.

Objectives

The objectives of this report are to:

- Explain the mission, dredging environment, problems, and issues concerning the management of the four hopper dredges in the Corps minimum dredge fleet.
- Present the actions implemented to improve the hopper dredge data and information.
- Describe the results of analyses of costs and benefits of existing and proposed restrictions on the uses of Corps hopper dredges.
- Propose and evaluate options.
- Recommend the appropriate investment decisions regarding Corps hopper dredges.

Mission

The Corps navigation mission is to provide safe, reliable, efficient, effective and environmentally sustainable waterborne transportation systems for movement of commerce and national security needs. The program includes 25, 000 miles of commercially navigable channels, 627 shallow harbors, and 299 deep-draft harbors. The navigation program is a vital link to waterborne commerce and handles much of the nation's international trade (2.4 billion tons of commerce, of which 1.1 billion tons are foreign commerce).

Data

The Dredging Information System (DIS) was used as the data source for Corps dredging contracts. Several discrepancies were found and a data call was initiated to have the data corrected. The DIS has been modified to contain the data fields considered relevant, important and useful for the management of the Corps dredging program. On 17 February 2004, the modified DIS was put into production. Some of the pick lists have been expanded and two new data fields have been added to the selection screens. The data has been screened, reviewed and accepted as correct by all Corps of Engineer districts and by the dredging industry.

Requirement

While the volume of dredged material varies annually because of natural forces and man-made limits, the Federal dredging workload has averaged 266 million cubic yards, from FY 1993 to FY 2004, as shown in Table 1.

Almost 30 percent (66 MCY) of the total dredging performed in FY 2003 was hopper work, mostly done in ocean entrances of large harbors and large rivers. The workload varies by region, with nearly 80 percent of the workload in the Atlantic and Gulf regions. This regional workload also varies throughout the year. Projected additional hopper dredging requirements associated with major deepening projects underway and planned throughout the United States will influence regional demands on the hopper dredge fleet. Non-Federal hopper dredge requirements also represent an impact on workload demands. There may be some adjustment in requirements that results in increased workload, but the additional 18% increase in industry capability with the November launching of the 12,000 cubic yard Manson hopper dredge will most likely be able to cover additional capability requirements.

Capability

The Corps contracts for the services of the U.S. private industry and uses its Corps- owned hopper dredges to accomplish the workload. The size and capability of each hopper dredge must be considered when assigning or procuring work. For example, a large hopper dredge, due to its size and draft will not be able to do work in a shallow-draft project. Conversely, a small hopper dredge may not be an efficient and effective choice for a deep-draft project with a long haul distance to a disposal site.

Table 1. Corps and Industry Dredging History

Fiscal Year	Dollars (millions)			Quantity (million cubic yards)		
	Maint	New	Total	Maint	New	Total
1993	$410.23	$104.66	$514.88	235.51	33.50	269.00
1994	$426.71	$100.82	$527.53	264.71	36.96	301.67
1995	$408.19	$122.84	$531.03	217.13	33.99	251.12
1996	$425.02	$89.74	$514.77	234.28	24.37	258.64
1997	$494.45	$127.48	$621.93	252.74	32.18	284.93
1998	$532.50	$178.00	$710.50	211.30	27.30	238.60
1999	$580.10	$232.20	$812.30	241.70	42.10	283.90
2000	$541.00	$280.70	$821.70	226.70	58.60	285.30
2001	$557.00	$310.70	$867.80	217.60	50.80	268.50
2002	$558.70	$364.20	$922.90	204.50	44.10	248.60
2003	$597.20	$290.10	$887.30	191.00	42.80	233.80
2004	$617.10	$284.52	$903.13	222.88	51.13	274.24

Source: Corps Continuing Cost data.

Table 2. Dredging Workload FY03

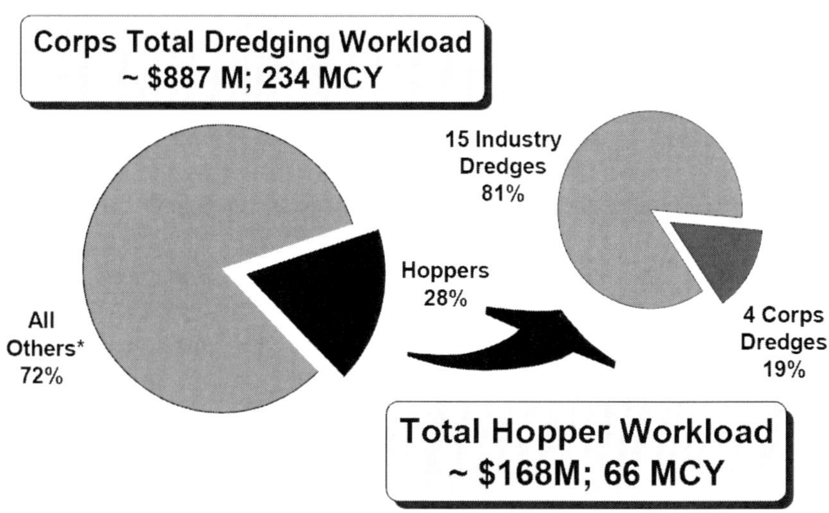

*Bucket, pipeline, dustpan, sidecaster, special purpose.

Industry accomplishes more than 80 percent of the hopper dredge workload. Currently, five companies own the 15 industry hopper dredges used by the Corps, as shown in Table 3. Manson Construction Company has commenced construction of a 12,000 cubic yard hopper dredge, the *Glenn Edwards*, that will be ready in November 2005, thus increasing industry capacity by 18%.

Corps-Owned Fleet

The four Corps hopper dredges were designed and built as ocean going dredges. The *Essayons*, *Wheeler*, and *Yaquina* were launched in the early 1980s, and the *McFarland* was launched in 1966.

While the table above reflects separate columns for the Corps and industry hopper dredges, the ability to accomplish the dredging requirements of the navigation mission is dependent upon the operation of the entire fleet of hopper dredges, as described above.

Table 3. Total Hopper Dredge Fleet - 19 Hopper Dredges (15 Industry Dredges/5 Companies; 4 Corps Dredges)

Class	Industry, Owner	Corps, District
LARGE HOPPER (6000 – 12,000 cy)	*Liberty Island,* Great Lakes *Stuyvesant,* Bean/Stuyvesant *Eagle 1,* Bean/Stuyvesant	*Wheeler,* New Orleans *Essayons,* Portland
MEDIUM HOPPER (3000 – 6000 cy)	*Bayport*, Manson *Columbia,* B+B *B.E. Lindholm,* Weeks Marine *R.N. Weeks,* Weeks Marine *Newport,* Manson *Dodge Island,* Great Lakes *Manhattan Island,* Great Lakes *Padre Island,* Great lakes *Sugar Island,* Great Lakes	*McFarland,* Philadelphia
SMALL HOPPER (0-3000cy)	*Northerly Island,* Great Lakes *Westport,* Manson *Atchafalaya, B+B*	*Yaquina,* Portland

Previous Restrictions on Corps Hopper Dredges

Prior to Fiscal Year 1992, the four Corps hopper dredges worked approximately 230 days each. Beginning in Fiscal Year 1993 restrictions were placed on the operation of the Corps hopper dredges.

(A) The Energy and Water Development Appropriations Act, 1993, required the Corps to advertise for competitive bid at least 7.5 million cubic yards of the hopper dredge volume accomplished with Government-owned dredges in Fiscal Year 1992. This requirement was included in each subsequent Appropriations bill, through the FY 1997 Act, and industry demonstrated they could perform in a timely manner and at a reasonable cost, as evidenced by the bidding results and contract performance. The Corps subsequently acknowledged Congress' intent and reduced the Government hopper dredges annual operations to 180 dredging days.

(B) Section 237 of the Water Resources Development Act of 1996, directed the Secretary of the Army to initiate a program to increase the use of private-industry hopper dredges for the construction and maintenance of Federal navigation channels. In order to carry out this requirement, the following actions were required:

- Place the Federal hopper dredge *Wheeler* in ready-reserve status effective 1October 1997.
- Develop and implement procedures to ensure private-industry hopper dredge capacity is available to meet routine and time-sensitive dredging needs.
- Limit active Federal hopper dredges to no more work than the average assigned in the past three fiscal years, and no less availability and utilization than that which occurred in fiscal year 1996.

The *Wheeler* has been in ready reserve since 1 October 1997, and has been used in conjunction with the combined industry and Corps hopper dredge fleet during peak workload conditions. The implementation of the Industry–Corps Hopper Dredge Management Group (ICHDMG) has successfully demonstrated the ability to manage the dredging requirements with the existing Corps and industry hopper dredge fleet. The ICHDMG is a chartered operations working group comprised of all hopper dredge using districts and divisions and members from all five hopper dredge owning companies. Frequent meetings and weekly tracking of all hopper dredges ensure timely resolution of issues and consistent knowledge of hopper dredge capability. The ICHDMG has established a formal notification and decision process for time-sensitive, urgent, and emergency dredging requirements. This process ensures that industry has an opportunity to perform this critical work, and the navigation mission requirements are accomplished effectively.

(C) The Energy and Water Development Appropriations Act of 2002, further limited the Corps hopper dredge *McFarland* to 85 days of work in the Delaware River. Most of the work in the Delaware River was removing spot shoals in areas requiring pumpout of the dredged material in upland disposal sites. This work has historically been performed by bucket dredges or pipeline dredges. Subsequent appropriations acts did not contain the same restrictions, and in Fiscal Year 2004, the Corps scheduled 140 days of hopper dredging work for the dredge.

Environmental Constraints and Other Considerations

All dredging operations must comply with environmental laws and regulations and must be consistent with applicable state water quality and coastal management criteria. Many navigation projects have specific environmental windows during which dredging cannot be performed.

There are also other environmental restrictions governing the Atlantic, Pacific, and Gulf Coasts, the Great Lakes and inland waterways. Because of concern for endangered and threatened species of sea turtles, the Corps agreed to limit hopper dredging in the Southeast to a window of 4 months – December through March. The concern for right whales, sturgeon, and migrating fish imposed additional restrictions including limiting dredging to certain windows and reducing dredge operating speeds. Other inefficiencies are created when hopper overflow is restricted or when hauling or pumping distances are increased to reach an environmentally acceptable disposal site.

Environmental dredging windows require the Corps to use industry or Corps dredges at only certain times of the year. In some cases, competition is limited because only certain dredges meet the environmental considerations. Ongoing environmental negotiations indicate that the number of resulting operational restraints will continue to increase.

There are several other considerations. Weather plays a significant role as to when dredging can be done. There are also economic considerations. Beach nourishment work is normally done in the winter to prepare for the summer tourist season. Sometimes there are legal considerations, where litigation may dictate when dredging can or cannot be done. Under these conditions, managing the national dredging assets to meet national and regional requirements is increasingly challenging. The net effect of these constraints is the requirement to accomplish hopper dredging in several projects all at the same time. This has resulted in peak workload demands on the entire hopper dredge fleet and can significantly impact hopper dredge availability.

Current Conditions

In Fiscal Year 2003, 66 million cubic yards were dredged by the combined Corps and industry fleet of 19 hopper dredges at a cost of $168 million. The Corps performed approximately 19% of the work by volume, dredging 506 days with its 4 hopper dredges, and five private contractors performed the other 81% with 15 hopper dredges. One of the Corps hopper dredges, the *Wheeler*, in a ready reserve status, only dredged 58 days. The Corps hopper dredge *McFarland* was limited to 85 days of work in the Delaware River as a result of the 2002 Energy and Water Development Appropriations Act. The other two Corps hopper dredges, the *Essayons* and *Yaquina*, worked 185 and 178 days respectively. Industry hopper dredges worked a combined total of 3581 days during Fiscal Year 2003. For Fiscal Year 2004, the industry worked a total of 2842 days, which is 79 percent of the FY 2003 workload. This low utilization is due to unusually reduced dredging requirements and constrained budgets. Most of this work occurred during the winter months, and many of the hopper dredges were subsequently idle. The Corps hopper dredges worked 565 days, with the *Wheeler* working 55 days, the *Essayons* working 192 days, the *Yaquina* working 178 days, and the *McFarland* working 140 days. The industry workload in the table below includes Federal maintenance and new work, beach nourishment, and private work.

Table 4. Industry Workload Days Worked

Dredge	FY 00	FY01	FY02	FY 03	FY 04
Atchafalaya	283	278	205	250	207
Bayport	249	167	352	286	206
Columbia	293	281	263	278	321
Dodge Island	308	310	199	262	108
Eagle	351	283	324	324	279
Liberty Island	--	--	153	197	167
Lindholm	217	304	246	290	209
Manhattan Island	279	343	223	173	171
Newport	143	230	272	252	222
Northerly Island	208	67	85	150	0
Padre Island	133	329	295	313	116
R.N. Weeks	250	275	279	280	251
Sugar Island	305	287	317	208	181
Stuyvesant	279	302	235	236	293
Westport	122	67	191	82	111
Total Days	3420	3523	3639	3581	2842

Source: Data submitted by 5 hopper dredging companies.

CORPS HOPPER DREDGE CONCERNS

The Issue

There is concern regarding the existing and proposed restrictions on the use of the Corps hopper dredge fleet.

The Problem

The composition of the minimum fleet is based on four provisions of Public Law 95-269, enacted 26 August 1978.

Table 5. Excerpts from Public Law 95-269 (1978)

"The Secretary shall have dredging and related work done by contract if he determines private industry has the capability to do such work and it can be done at reasonable prices in a timely manner."
"As private industry reasonably demonstrates its capability...to perform the work done by the federally owned fleet, at reasonable prices and in a timely manner, the federally owned fleet shall be reduced in an orderly manner, as determined by the Secretary, by retirement of plant."
To carry out emergency and national defense work the Secretary shall retain only the minimum federally owned fleet capable of performing such work and he may exempt...such amount of work as he determines to be reasonably necessary to keep such fleet fully operational."
"...the Secretary may retain so much of the federally owned fleet as he determines necessary, for so long as he determines necessary, to insure the capability of the Federal Government and private industry together to carry out projects for improvements of rivers and harbors."

PL 95-269

The first provision directs the Corps to use private industry when it has the capability to do the work at reasonable prices and in a timely manner. The second provision directs the Corps to retire dredges when industry demonstrates capability, timeliness, and reasonable prices. The third provision indicates that the Corps may retain only the minimum federally owned fleet to carry out emergency and national defense work. The Corps may set aside work as reasonably necessary to keep this minimum fleet fully operational. The fourth provision indicates that the Corps may retain enough of the federally owned fleet to ensure sufficient capability of the combined Federal and private industry to carry out the workload. These four provisions are used as the basis and focus for the analysis.

Industry Hopper Dredge Fleet

Encouraged by Congress in 1978 with the passage of PL 95-269, industry has built 15 hopper dredges to perform work formerly done exclusively by the Government fleet. The policy established by PL 95-269 has been supported by every succeeding Administration and Congress through legislative and regulatory actions, including WRDA 92 and WRDA 96. With this Congressional incentive, industry has responded as desired, increasing investment and adding capacity in significant measure. Appendix C, Table C-1, displays the dates of introduction of each hopper dredge in the fleet.

Limited Number of Hopper Dredges

There is a limited number (15) of hopper dredges owned by 5 companies in private industry to perform work by contract. A limited resource (capability) impacts timeliness and costs in an open marketplace. For example, there is an inverse relationship between the number of bidders and contract cost. The greater the number of bidders, the more likely that the contract will be awarded at or below the Government cost estimate. Single-bid contract awards are likely to be above the Government estimate. Over the last four fiscal years, 2001 through 2004, the winning bid as a percentage of the Corps estimate for the most frequent type of single-bid projects, those on the Mississippi River, have been 8.5 percent above the Government estimate. The following table lists the industry hopper dredging contract bidding information from FY 1990 through FY 2004. The table compares the total of all winning bids with the total of the Government Estimates for each year, and shows the average number of bids per contract. Since fiscal year 1998 when the *Wheeler* went into ready reserve, the average winning bid as a percentage of the Government estimate has been below the Government estimate.

Hopper dredge workload varies from month to month and peak workload demands can occur in any month depending upon the requirements. Generally, the combined industry-Corps hopper fleet has been able to meet the workload. However, industry alone has not been able to meet peak demands. This is especially a concern when an industry dredge is committed to a long-term contract, when a dredge does not have capability (e.g., the dredge requires more depth that the authorized channel) for the job, or in the case of the West Coast where there may only be one industry hopper dredge available.

The peaks on this graph can vary from month to month and recently seem to be more concentrated in the winter months. Two major factors are the winter sea turtle dredging window in the Southeast and the high water season in the lower Mississippi River. When these two coincide, all available hopper dredging capacity may be required. This table demonstrates that the requirement to accomplish hopper dredging in several projects at the same time results in peak workload demands on the entire hopper dredge fleet that significantly impact hopper dredge availability.

Table 6.

Awarded Contracts FY 1990 - 2004 For Hopper Dredging						
FY	**Awarded Cubic yards**	**Average No bids**	**Total of Winning Bids**	**Total of Government Estimate**	**Percent of Government Estimate**	**Percent of Government Allowable ***
1990	52,310,270	3.0	$72,508,932	$80,582,866	90%	72%
1991	42,786,187	3.2	$42,283,989	$54,589,333	77%	62%
1992	33,953,905	2.6	$45,892,132	$54,588,397	84%	67%
1993	47,021,989	2.6	$72,262,468	$80,966,780	89%	71%
1994	63,002,356	2.3	$138,038,626	$157,804,394	87%	70%
1995	43,388,000	2.4	$92,925,222	$88,333,538	105%	84%
1996	46,009,910	2.7	$111,834,196	$123,976,115	90%	72%
1997	62,399,556	2.1	$107,360,401	$103,493,270	104%	83%
1998	61,044,341	1.9	$115,337,432	$116,394,700	99%	79%
1999	64,293,329	2.3	$112,224,650	$115,571,760	97%	78%
2000	40,675,200	3.1	$79,574,628	$105,537,766	75%	60%
2001	58,633,532	3.1	$94,839,621	$118,194,152	80%	64%
2002	70,986,830	2.6	$146,167,793	$155,148,478	94%	75%
2003	61,282,300	2.4	$89,491,574	$102,559,105	87%	70%
2004	38,551,000	3.9	$40,210,195	$64,208,034	63%	50%

* Award can be 125% of Government Estimate.

Source: DIS.

Table 7. Hopper Dredge Workload

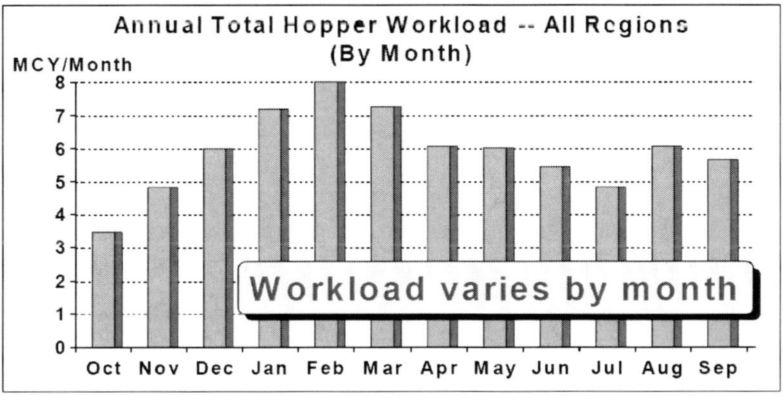

CURRENT OPERATIONAL PRACTICE

Adjust Operating Days

In order to comply with existing legislation as discussed earlier, the Corps is currently operating with reduced schedules. Further reductions would offer additional workload to industry, which would encourage investment, and would increase the daily operating rates of the Corps hopper dredges. Increasing the scheduled work for the Corps dredges would reduce the daily rates, however, industry utilization would be reduced, with resulting unrecoverable fixed costs, increased operating costs, potential loss of capacity, and reduced competition

Ready Reserve

The *Wheeler* was placed in ready reserve status on October 1, 1997, in accordance with Public Law 104-303, Water Resources Development Act of 1996 (WRDA 96), Section 237. As specified in WRDA 96, one of two circumstances must exist before the *Wheeler* may be activated from ready reserve status for an urgent dredging assignment:

1. Private industry failed to submit a responsive and responsible bid for advertised work.
2. A private industry contractor failed to comply with contract specifications.

WRDA 96 states that the Secretary may place the *Wheeler* in active status when either of the conditions described in the previous section are satisfied. The procedure for activation of the *Wheeler* uses notification and feedback from the Industry–Corps Hopper Dredge Management Group (ICHDMG), a group of Corps and private industry hopper dredge managers and representatives. This information is used as input to a formal "Raise the Flag" procedure for hopper dredge management. This procedure includes the following:

a) When a Corps district recognizes a possible future need for a hopper dredge, a "Yellow Flag" notice is distributed via electronic mail to all ICHDMG recipients.
b) During the "Yellow Flag" phase, the ICHDMG members may provide their input to Corps Headquarters regarding possible alternative solutions for supplying a hopper dredge or for ready reserve activation of the *Wheeler*.
c) If the District's need for hopper dredging services becomes imminent, the "Red Flag" notice is distributed via electronic mail to all ICHDMG recipients.
d) After distribution of the "Red Flag" notice, the ICHDMG members are again given the opportunity to provide input to Corps Headquarters.
e) If no alternative solutions are chosen, and the WRDA 96 conditions for activation have been satisfied, Corps Headquarters will activate the *Wheeler*.

This concept implies that any Corps dredge in a ready reserve status would be at the dock awaiting assignment, except during periodic training days. The *Wheeler* schedules 55 training days per year performing dredging in projects. An appropriate analogy is a community fire engine awaiting the bell. An advantage of this concept is that the Corps would be prepared to meet peak requirements and unforeseen situations. All industry dredges are currently under basic ordering agreements known as the Corps of Engineers Reserve

Fleet (CERF). Under CERF, the Government can obtain the use of an industry dredge working on a non-urgent project to work on an urgent project. On the other hand, this concept presents some financial difficulties. If the dredge is sitting at the dock for a large portion of the year, a subsidy is required to fund the costs of crew and to be fully ready to respond. In the case of the *Wheeler*, this subsidy is $8 million per year. One alternative would be for the industry to provide a ready reserve dredge and seek to arrange for a less expensive reserve capability.

ASSUMPTIONS

Strategic Ports

There are 14 designated Strategic Ports in the United States, all of which are Federally authorized deep-draft navigation projects. During mobilization operations, these ports are given highest priority to be maintained to support overseas transportation requirements. Monitoring channel conditions and dredging requirements of Strategic Ports has always been assumed to be an integral part of Corps operations. Procedures developed in the Industry Corps Hopper Dredge Management Group (ICHDMG) are in place to rapidly respond to any unforeseen requirements that may occur. Consideration of capability requirements for these Strategic Ports has been included in the risk assessment for the ports, harbors and navigation projects.

Contingency Operations

The Corps Government hopper dredge fleet represents a national asset that could be mobilized to support military contingency operations. However, in recent years there has been discussion on the use of the government hopper dredges for this purpose including for potential use in military operations at the Port of Umm Qasr in Iraq. In this instance, foreign and domestic capability from the private sector were more than sufficient to meet operational requirements within the theater of operations. In this instance, there was a great reluctance to dispatch Government hopper dredges critical to dredging in the lower Mississippi River or the Northwest abroad to support contingency operations.

This report does not assume potential use of the government hopper dredge fleet in direct support of contingency operations based upon foreign and American owned hopper dredge capabilities, political considerations of removing the hopper dredge fleet from domestic waters in their support to the civil works mission, and the historical experience of the existing hopper dredge fleet never participating in such operations. Should the Government hopper dredge fleet be utilized it would be in U.S. waters in backfilling private dredges deployed overseas. However, the management practice would be for the Corps to work with the Department of Defense, private dredging industry, and the Congress to ensure sufficient dredging capability was available to perform the civil works mission. Therefore, contingency operations are not considered to be an issue that would affect the size, configuration, or utilization of the Government's hopper dredge fleet.

Worst Case Scenario

A worst case scenario was developed to evaluate options and help put risk factors in proper perspective. It was not necessary to create a hypothetical version, because circumstances that occurred in late Fiscal Year 2004 and early Fiscal Year 2005 offered a real worst case condition that substantially challenged the industry and Corps hopper dredge fleet. The scenario began with the onslaught of four consecutive hurricanes striking Florida, causing substantial shoaling in several major navigation projects and severe beach erosion in a number of localities. As surveys were completed and the scope of the hurricanes' impacts were realized, emergency procurements were issued for beach renourishment work and dredging navigation channels that were best accomplished by hopper dredges. At this same time, late fall of 2004, the river stage on the Mississippi River below New Orleans began to rise, resulting in increased shoaling in Southwest Pass, requiring immediate response by hopper dredges. The winter hopper dredging window, December through March, for dredging Federal projects in the Southeast Atlantic affected by endangered and threatened sea turtles, was about to open. By mid-January 2005, all industry hopper dredges on the Atlantic and Gulf Coasts were engaged (except the small hopper *Northerly Island*), the *McFarland* was in the shipyard, and the *Wheeler* was working in Southwest Pass. There was additional emergency beach work best suited for hopper dredges that needed to be accomplished, but was able to be deferred without being vulnerable to environmental window issues. During this peak workload scenario, the largest industry hopper dredge, the *Stuyvesant*, experienced engine trouble and had to stop work, creating a capability shortfall. A series of adjustments resulted, including subcontracting to other industry hopper dredge owners to complete critical work, and some delays in starting other work. Subsequent to this event, increased shoaling in Mobile Harbor created the need for an additional hopper dredge, resulting in calling out the Wheeler, as the McFarland was also fully engaged in work in the Gulf. While there were additional procurement and operational issues, this synopsis of a worst case scenario adequately describes an extreme case as a result of four consecutive hurricanes, high water in the Mississippi River, substantial shoaling in various navigation channels, the need to dredge South Atlantic ports during the winter turtle dredging window, and the operational vulnerabilities of fully utilized equipment. However, the likelihood of such a scenario occurring must be balanced with the evaluation of what is a reasonable capability to respond to unforeseen peak workload requirements. It is also necessary to evaluate the sequential response capability afforded by having a ready reserve capability and a flexible industry capability. The likelihood of this worst-case scenario occurring is relatively low. Having four hurricanes in a row with the extent and magnitude of damages experienced is not a common occurrence. The ability to use the combined hopper dredge fleet to respond to this extraordinary workload was helped by the reserve capability of the Wheeler, and its ability to rapidly respond when the need arose. Accordingly, the industry demonstrated that they could effectively respond, and had the flexibility to adjust to ever changing requirements and challenges.

Combined Corps Hopper Dredge Fleet

Unfortunately, there are competing concerns among the many interests involved. Most of the concerns can be grouped into three areas – the dredging requirement, the capability to satisfy the requirement, and the cost of doing business. The industry has 15 hopper dredges and seeks to fully utilize these high cost vessels to ensure recouping their fixed costs. If existing policy underlying almost thirty years of legislation were dismissed and all restrictions were removed from the Corps hopper dredges, then these four dredges would be scheduled first and industry would get the remaining work. If the Corps dredges were fully utilized or even heavily utilized, they would offer no resources for peak or unforeseen demand. Industry could be placed in the position of having to supply all peak and unforeseen demand, and if there were low demand, industry would have to absorb the impact.

Table 8. Potential Impacts to Industry Fixed Costs

Corps ops	Status quo* 503 days	FY04** 565 days	180 days/ dredge 720 days	Maximum*** 885 days
Industry lost Fixed costs (3581 days FY03)	0%	-2%	-6%	-11%
Industry lost Fixed costs (2842 days FY04)	-19%	-21%	-25%	-30%

*Wheeler 55 days, McFarland 85 days, Yaquina 178 days, Essayons 185 days.
**Wheeler 55 days, McFarland 140 days, Yaquina 178 days, Essayons 192 days.
*** Wheeler 245 days, McFarland 230 days, Yaquina 195 days, Essayons 215 days.

The following table shows the relationship of Corps scheduled work under various options and the potential impact on industry's ability to recoup their fixed costs.

If all restrictions were removed and all four Corps hopper dredges worked the maximum possible time and the dredging requirements are as low as FY04, it is assumed that industry would potentially not be able to recoup 30 percent of their fixed costs, which could result in a 30 percent reduction in plant. The table above shows the potential percentage impact to industry's ability to recoup fixed costs using the FY 03 total days of 3581 as the basis. As the Corps scheduled workload increases, the added Corps days are subtracted from the 3581 days to derive a percentage decrease. For the abnormally low utilization realized in FY 04, the percentage is computed against the FY 03 basis of 3581 days. Initially industry would seek to increase their costs to recoup additional costs and may minimize maintenance expenditures of operating dredges. However, if this trend continued, industry may be forced to either seek additional work outside the United States, which could jeopardize their ability to return because of Jones Act requirements, or they may have to reduce the number of dredges in their fleets. Either action would result in loss of capacity. There could be a significant reduction in private plant, particularly if the Corps should increase use of its own fleet. When peak workload demands did occur, subsequent to this reduced capacity, there would not be enough capacity to respond to the requirements. This is a qualitative analysis and does not represent actual industry fixed costs. The table reflects relative impacts to the industry based on an assumption that FY03 workloads would be the basis for recouping all fixed costs. The industry data used in this report is not used to make conclusions, but is used for relative comparisons for analyses of workload and cost impacts. To verify the industry data would require extensive auditing and is beyond the scope of this report.

Benefits of Corps Hopper Dredges

The Corps hopper dredges are not only critical to the accomplishment of the hopper dredging requirements of the Navigation Program but serve other important roles as well. First, they serve as the Nation's hopper dredge response fleet, ready to respond to emergency and National Defense needs. In this role, they are the marine transportation fire engine awaiting the bell. Second, they serve as a source of knowledge and expertise for the Corps of Engineers, ensuring that the Corps has the depth of understanding and technical expertise to negotiate and manage the industry hopper dredge operations and contracts. In this role, the operation of the Corps hopper dredges serve to provide the knowledge base of Corps operations managers beyond just the textbooks and guidance documents. Third, the Corps hopper dredges serve to ensure that costs will be reasonable. With such a limited number of vessels in the fleet, and during peak workload periods when only one bidder may be available, there is a tendency to exercise the principles of supply and demand, and costs will rise. The Corps presence will serve as a deterrent for potential cost increases. A current example is the Wheeler being called out in February 2005, to perform work in the Mississippi River when a single industry bid exceeded the awardable amount.

OPTIONS

The following 12 options were considered in evaluating the configuration of the Corps hopper dredges: Status Quo (FY03); Maximum Use of All 4 Corps Dredges; Operating the *McFarland* 140 days, the *Essayons* 215 days, the *Yaquina* 195 days and the *Wheeler* in Ready Reserve; Retiring the *McFarland*, operating the *Wheeler* 180 days, the *Essayons* 215 days, and the *Yaquina* 178 days; Retiring the *McFarland*, operating the *Wheeler*, *Essayons* and *Yaquina* the maximum days; Operating the *Wheeler* and *McFarland* 140 days, the *Essayons* 185 days, and the *Yaquina* 178 days; *McFarland* and *Wheeler* in Ready Reserve, operating the *Essayons* 215 days, and the *Yaquina* 178 days; Retire the *McFarland*, *Wheeler* in Ready Reserve, operating the

Essayons 215 days, and the *Yaquina* 178 days; Retire the *McFarland*, place other 3 Corps dredge in ready reserve; Retire all 4 Corps dredges; Retire the *McFarland*, *Wheeler* and *Yaquina* in Ready reserve, operating the *Essayons* 215 days; and Retire the *McFarland*, *Wheele*r in Ready Reserve, operating the *Essayons* 185 days, and the *Yaquina* 178 days.

Definitions

Terms used in the Options summary table below are defined and explanations of the one-time costs and assumptions are addressed below.

Retire

The Corps dredge would be declared excess to the Corps needs and retired, at a considerable financial loss of sunk costs. After cleaning and decontaminating the dredge, the Corps would relinquish ownership. If no other Government agency had a need for the vessel, it would probably be sold as scrap or sold overseas to partially offset sunk costs. The Water Resources Development Act of 1986 prohibits excess Corps dredges from being used in the United States as dredges.

Costs of Options

Costs of alternatives can be grouped into three categories: costs associated with the disposition of the Corps dredge (one time cost), costs associated with any change in operation of the Corps hopper dredges (change in status, operating days), and costs associated with contracting work previously performed by Corps hopper dredges or no longer performed by industry. The contracting costs reflect the change in cost for the respective option as compared with Status Quo.

Several options were considered to evaluate the cost of fully operating the Corps dredges and other scenarios with various restrictions in scheduled working days. The costs do not reflect the long-term costs to the industry for those options that substantially reduce the utilization of the industry hopper dredges. Such costs could only be attained by audit of the industry. Any option that includes the use of the Corps hopper dredge *McFarland* must include the rehabilitation and repowering costs estimated to be $20 million. Any option that reflects retiring the *McFarland* includes one-time costs of $22 million for separation of crew costs, clean-up, and pay off of the remaining corpus.

One Time Costs

Options proposing retiring Corps dredges show the one time costs in parentheses. These costs include preparing the vessel for disposition, and include cleaning and decontamination, and the costs associated with crew separation, including retraining, change in station, reduction-in-force benefits, payoff of the remaining corpus and administration.

Options Summary

The following table is a summary of all 12 options. All of the options are compared against the costs reflected for the Status Quo (FY03) Option. The Total Corps Work Days is the annual accumulation of all dredging days performed, or proposed, by all 4 Corps hopper dredges. The Government cost is the total annual cost to operate the 4 Corps hopper dredges, including any subsidies. One-time costs are shown in parentheses. The Contracting cost for the Status Quo Option is the actual cost of all industry hopper dredge work performed on Corps projects in FY 03, and is the estimated total cost for the other options. The Total cost is the sum of the Government annual costs and the Contracting costs. For a detailed analysis of each option see Appendix A. Appendix B details the costs associated with each option.

Evaluation of Options

In evaluating the various options, the changes in daily operating costs become a significant factor in developing the best business case for the configuration of Corps hopper dredges. The following discussion regarding daily rates and comparisons with industry daily rates is a preamble to the evaluation of the options. The analysis will consider costs, risks to industry viability, and risks to the ability of the Corps to be responsive to the needs of ports, users and stakeholders of the Federal projects.

Table 9. Options Summary

Option	Total Corps Work days	Costs		
		Government	Contracting	Total
Status Quo: 55 days for *Wheeler*, 85 days for *McFarland*, 178 days for *Yaquina* and 185 days for *Essayons*	503	$49.1M ($20 M rehab for *McFarland*)	$118.7 M	$167.8 M
Opt 1: Maximum days for all 4 Corps dredges *Wheeler* 245 days, *McFarland* 230 days, *Essayons* 215 days, *Yaquina* 195 days	885	$56.7M ($20 M rehab *McFarland*)	$100.5M	$157.2 M
Opt 2: 140 days *McFarland*, 55 days for *Wheeler*, 178 days for *Yaquina*, 192 days for the *Essayons*	565	$47.8M ($20 M rehab *McFarland*)	$116.8 M	$164.6 M
Opt. 3: Retire *McFarland*, 180 days for *Wheeler*, 215 days for *Essayons*, 178 days for Yaquina	563	$40.7M ($22 M retirement costs)	$112.6M	$153.3 M
Opt 4: Retire McFarland, Wheeler 245 days, Essayons 215 days, Yaquina 195 days	655	$41.9M ($22 M retirement costs)	$107.7M	$149.6 M
Opt 5: McFarland& Wheeler 140 days, Essayons 185 days, Yaquina 178 days	643	$50.7M ($20 M rehab McFarland)	$111.2M	$161.9 M
Opt 6: McFarland& Wheeler ready reserve, Essayons 215 days, Yaquina 178 days	533	$49.5M ($20 M rehab McFarland	$118.7M	$168.2 M
Opt 7: Retire McFarland, Wheeler ready reserve,	448	$37.0M ($22 M retirement costs)	$120.8M	$157.7 M

Option	Total Corps Work days	Costs		
		Government	Contracting	Total
Essayons 215 days, Yaquina 178 days				
Opt 8: Retire McFarland, Wheeler, Essayons, & Yaquina in ready reserve	165	$36.6M ($22 M retirement costs)	$132.9M	$169.5 M
Opt 9: Retire McFarland, Wheeler, Essayons, and Yaquina	0	$0M ($178 M retirement costs)	$142.6M	$142.6 M
Opt 10: Retire McFarland, Wheeler and Yaquina in ready reserve, Essayons 215 days,	325	$36.6M ($22 M retirement costs)	$125.2M	$161.8 M
Opt 11: Retire McFarland, Wheeler in ready reserve, Essayons 185 days, and Yaquina 178 days	418	$36.6M ($22 M retirement costs)	$120.8M	$157.4 M

Corps Daily Rates

Changing the scheduled days for the Corps hopper dredges affects the daily rate accordingly. The figure below describes the differences in daily rates for the options. The *Wheeler* daily rate in ready reserve is held at $75,000 to minimize impacts to those projects where the dredge performs its training. Accordingly, the *McFarland* daily rate operating under the 85 day restriction had its daily rate held at $65,000 per day, and required a $7 million subsidy to fund being tied at the dock. All other daily rates are computed based on total costs divided by the number of scheduled days. The 140-day scenario for the *McFarland* and the *Wheeler* represents the maximum days the dredges can work with a reduced crew and the minimum days the dredge can work without requiring a subsidy.

Table 10. Impact of Options on Daily Rate

Dredge	Status Quo	Maximum Days Option 1	Other Option	Other option
Wheeler	$75,000 for 55 days ($8 million subsidy) $12.3 million	$68,415 for 245 days $16.8 million	$88,150 for 180 days $15.9 million	$99,500 for 140 days $ 13.9 million
Essayons	$87,000 for 185 days $16.1 million	$76,607 for 215 days $16.5 million	$84,304 for 192 days $16.2 million	$87,000 for 55 days ($11.3 million subsidy) $16 million
Yaquina	$47,000 for 178 days $8.4 million	$44,655 for 195 days $8.7 million	none	$47,000 for 55 days ($5.8 million subsidy) $8.3 million
McFarland	$65,000 for 85 days ($7 million subsidy) $12.5 million	$64,000 for 230 days $14.7 million	$87,925 for 140 days $12.3 million	

Industry Daily Rates

In order to evaluate the options, and the cost impacts, the relationship between Corps daily rates and comparable industry dredge daily rates must be considered. This comparison is difficult and there is no absolute data set that can be used. The comparisons below are derived from information supplied by the industry and daily rental rates from previous bids from industry. Industry contracts normally include a separate bid item for mobilization and demobilization. This cost can range from $0 to several hundred thousand dollars, or more in the case of some West Coast bids. This cost would have to be applied to the total number of days the dredge worked to properly determine the daily rate of the industry dredge. However, a review of industry mobilization costs, exclusive of West Coast contracts, beach nourishment and new work, reveals that the addition to the industry daily rate for mobilization will average approximately $3,700 per day (based on an average of 52 contracts). For West Coast contracts, the average daily rate for mobilization is approximately $15,800 per day for a medium class dredge (based on an average of 7 contracts), however, this number is abnormally high and could substantially decrease with an industry presence on the West Coast. The industry rates in TABLE 11 include the additional average daily rate for mobilization.

If one just looks at the comparisons of daily rates between industry and the Corps, it is clear that similarly sized industry dredges can operate at less cost than Corps dredges. These rates reflect the current restricted operating days of Corps hopper dredges. However, there are other factors to consider such as capability, workload variability, and reduced work schedules of Corps vessels, and the fact that these dredges were not designed to be competitive with industry dredges. The Corps dredges were built to last longer periods of time, have much larger crew complements, and have different construction standards than commercial vessels. As an example, the cost differences between the McFarland and a comparable industry dredge can be attributed to labor costs (47 crew for McFarland and average 28 crew for industry), annual repair costs for the McFarland of approximately $5 million, higher depreciation rates, and the fact that crews are full time employees, where industry only employs most of their crew when working.

Analysis of Options

The various options discussed in this report indicate that selected changes in the number of Corps hopper dredges could result in some small annual savings. However, one of the options could reflect the greatest reduction in long-term capability, thus the greatest risk. The options and associated cost comparisons are not the only means of accomplishing the challenges mentioned above. The following discussion describes the potential risk of each option, including the Status Quo condition:

Status Quo

Status Quo offers the opportunity for industry to operate sufficient days to realize full funding of fixed costs. However, during a low shoaling year, such as FY04, industry could have as much as a 19 percent reduction in earnings. The Corps dredges would work the minimum number of days at 503 days. The *Wheeler* and *McFarland* would require subsidies of $8 million and $7 million respectively. The *McFarland* would require rehabilitation and modernization at an estimated cost of $20 million. The risk of not having available capacity is relatively low.

Option 1

Option 1 represents the greatest impact to the potential workload of the industry hopper dredges and may result in substantial long-term risk due to a loss of industry capability. Working the Corps hopper dredges the maximum possible days will result in an additional 382 days of scheduled work for Corps hopper dredges above status quo. The net reduction in costs is $10.7 million. Increasing the workload of the *Yaquina* beyond 178 days will result in this small class hopper dredge performing work normally best accomplished by more efficient medium class hopper dredges. There is a risk that scheduling the maximum amount of work for both the *Yaquina* and the *Essayons* will result in loss of industry capability on the West Coast. However, the additional days for the Essayons are not expected to impact industry workload, as anticipated longer haul distances to ocean disposal sites will add time to both industry and Corps operating requirements. Working all four Corps dredges the maximum days creates the scenario industry is most concerned about- the Corps schedules its dredges first, and industry gets what's left. Industry contends this option would result in substantial underutilization of their dredges, will result in some hopper dredges being tied up or scrapped, will create an inability to interest investors in building replacement dredges, and will cause a significant loss of ability to earn fixed costs for their dredges. As dredges are taken out of the market as a result of the workload decrease, the number of potential bidders will diminish, resulting in higher bids for work, as a result of loss of competition and higher costs for operation. If there are unforeseen dredging requirements, there will be little additional dredge availability to support a subcontract to ensure timely completion. This lack of ability to complete the work is a substantial risk to ports and maritime users. Because of environmental windows, there could be a potential for less than justified project dimensions for as much as 6 months. The *McFarland* would require rehabilitation and modernization at an estimated cost of $20 million. The risk of not having available capacity is relatively high.

Table 11. Comparison of Daily Rate between Industry and Corps

Dredge	Daily rate Status Quo	Daily rate of Comparable Industry Hopper dredge	Corps daily rate for maximum use (Option 1)
McFarland	$65,000	$34,905	$64,000
Wheeler	$75,000	$65,700	$68,415
Essayons	$87,000	$59,192	$76,607
Yaquina	$47,000	$35,800	$44,655

Option 2

Option 2 is similar to status quo, and reflects the FY 04 scheduled work for the Corps dredges, which includes operating the *McFarland* at 140 days to preclude the need for a $7 million subsidy. This option, if implemented, would require a $20 million investment in rehabilitating and modernizing the *McFarland.* The total days scheduled for Corps hopper dredges is 565 days and offers industry an opportunity to operate sufficient days to realize full funding of fixed costs. The net reduction in costs is $3.2 million. The risk of not having available capacity is low.

Option 3

The *Essayons* will schedule 215 days, and the *Yaquina* will continue to operate for 178 days. The *Wheeler* will be scheduled for 180 days and no longer require a subsidy. The *McFarland* will be retired. There will be some reduced capability as a result of retiring the *McFarland.* There will be one-time costs of $22 million to clean up and dispose of the vessel, pay off the remaining revolving fund corpus on the dredge, and for crew and employee separation costs. Corps dredges will work 563 days. The work performed by the *McFarland* is assumed to equate to 60 days by industry, but may not be all done by a hopper dredge. There will be 125 days of industry work added to the *Wheeler* scheduled work. These changes result in a $14.5 million reduction in costs. The additional days of scheduled work for the *Essayons* will not effectively reduce industry workdays on the West Coast. This option offers industry an opportunity to operate sufficient days to realize full funding of fixed costs. While there is a risk of lost capability in retiring the *McFarland,* the additional industry capability available and pending can accomplish the work. Retiring the *McFarland* will result in the loss of 47 dredge personnel and 30 district support personnel in the Philadelphia District. Scheduling the *Wheeler* for 180 days will substantially impact industry workload, and there will be no ready reserve capability to respond to peak workload demands. There will be some risk of not having available response capability.

Option 4

Option 4 would retire the *McFarland* and work the *Essayons*, *Wheeler* and *Yaquina* the maximum possible days, resulting in an additional 152 days of scheduled work for Corps hopper dredges above Status Quo. The net reduction in costs is $18.2 million. The $149.6 million total cost is one of the lowest cost options. There will be some reduced capability as a result of retiring the *McFarland.* There will be one-time costs of $22 million to clean up and dispose of the vessel, pay off the remaining corpus on the dredge and for crew and employee separation costs. The work performed by the *McFarland* is assumed to equate to 60 days by industry, but may not be done by a hopper dredge. There is a risk that scheduling the maximum amount of work for both the *Yaquina* and the *Essayons* will result in loss of industry capability on the West Coast. Increasing the workload of the *Yaquina* beyond 178 days will result in this small class hopper dredge performing work normally best accomplished by more efficient medium class hopper dredges. Scheduling the *Wheeler* for 245 days will substantially impact industry workload, and there will be no ready reserve capability to respond to peak workload demands. There will be some risk of not having available response capability.

Option 5

Option 5 would work the *Wheeler* and *McFarland* 140 days, the *Essayons* 185 days and the *Yaquina* 178 days, resulting in an additional 140 days of scheduled work for the Corps hopper dredges above status quo. The net reduction in total costs is $5.9 million. There would be no subsidies for Corps hopper dredges, however the daily rates for the *Wheeler* and *McFarland* would be $99,500 and $87,925 respectively. Operating the *Wheeler* and *McFarland* for 140 days can be done with reduced crews, and does not require any subsidy, however, the ability to respond to additional unforeseen requirements may be impacted by the reduced crew size. This could offer some risk to navigation projects.

Option 6

Option 6 would keep the *Wheeler* in ready reserve, place the *McFarland* in ready reserve, and operate the *Essayons* for 215 days and the Yaquina for 178 days. This option is very similar to Status Quo except the *Essayons* would work an additional 30 days. This is one of the highest cost options at $168.2 million. The parameters and risk factors are basically the same as Status Quo, and there would be $15 million subsidy required for the *Wheeler* and *McFarland.*

Option 7

Option 7 would keep the Wheeler in ready reserve, retire the *McFarland*, and operate the *Essayons* at 215 days and the *Yaquina* for 178 days, resulting in a reduction in the Corps workload of 55 days. There would be an $8 million subsidy for the *Wheeler* in ready reserve. The total annual costs would be $157.7 million, and there would be a one-time cost for retiring the *McFarland* of $22 million. This option results in very low Government costs at $37.0 million. The risk of not having available capacity is low.

Option 8

Option 8 retires the *McFarland* and places the other three Corps hopper dredges in ready reserve, resulting in a substantial reduction in Corps works to a total of 165 days. This will result in a subsidy for the three ready reserve Corps dredges of $25 million, and one time retirement costs for the *McFarland* of $22 million. This option will require two medium to large sized hopper dredges on the West Coast, plus at least one small class hopper dredge. This shift in dredge capability to the West Coast will offer moderate risk to navigation projects in the Gulf and Atlantic with the existing industry fleet, however, with the addition of the Manson hopper dredge and the three ready reserve Corps dredges, some risk is mitigated. There is a potential for less than acceptable competition for West Coast work, and costs could increase. This option has the highest total cost at $169.5 million.

Option 9

Option 9 is the most extreme option, retiring all four Corps hopper dredges, which results in one-time retirement costs of $178 million, and will substantially impact the Corps revolving fund income. The resultant total contract costs are $142.6 million, which does not adequately reflect the need for additional industry dredges to respond to peak workload demands. This option does not comply with the basic tenants of Public Law 95-269 to retain a Federal minimum fleet capability to ensure industry and the Corps together can carry out projects for improvements to rivers and harbors. This option reflects the highest risk to navigation projects, and the actual computed contract costs may not reflect true costs of not having any Corps hopper dredges.

Option 10

Option 10 operates the *Essayons* for 215 days, places the *Wheeler* and *Yaquina* in ready reserve and retires the *McFarland*, resulting in a reduction in the Corps hopper dredge workload of 178 days. There will be one-time retirement costs of $22 million for the *McFarland* and there will be a subsidy of $13.8 million for the *Wheeler* and *Yaquina* in ready reserve. The total cost of this option is $161.8 million. There is little risk associated with this option.

Option 11

Option 11 maintains the Status Quo and retires the *McFarland*. There will be one- time retirement costs of $22 million for the *McFarland* and there will be a subsidy of $8 million for the *Wheeler* in ready reserve. However, the net change in the cost of the program is a reduction of $10.4 million from the Status Quo, which covers the subsidy and increases work for industry. There is little risk associated with this option.

Table 12. Cost and Workload Comparison of Options

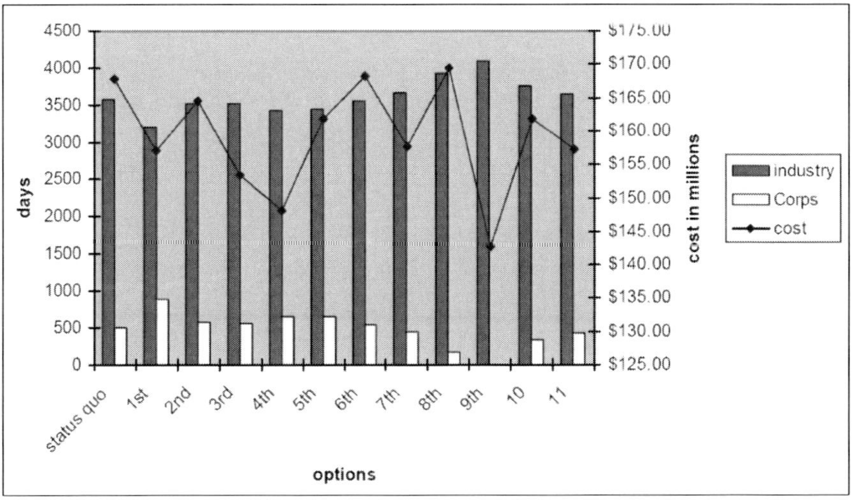

Minimizing Risk

If the options are considered in conjunction with management initiatives discussed in this report, there are opportunities for changes that may minimize the risk while improving the efficiency and effectiveness of the combined Corps and industry hopper dredge fleet. The Industry-Corps Hopper Dredge Management Group (ICHDMG) has identified initiatives that will help increase hopper dredge availability by reducing the number and duration of options in contracts, the development of a centralized scheduling and information system, and improved industry and Corps communication. A "Raise-the-Flag" procedure has been developed that ensures rapid information sharing, enhanced communication among industry and Corps districts and divisions, and the ability to make informed decisions regarding movement of hopper dredges for emergency response. This procedure has been successfully exercised several times. At any given time there is a high probability that at least two hopper dredges are capable of being mobilized to respond for a short period (no more than 30 days) to an emergency dredging requirement. This can be done with little impact to the projects from which the dredges are moved.

RECOMMENDATION

From the above discussion, the most reasonable option would be Option 7, which would schedule the *Essayons* for 215 days, operate the *Yaquina* for 178 days, keep the *Wheeler* in ready reserve, retire the *McFarland,* and continue with the initiatives from the ICHDMG. While there is a tendency to go to the least cost option, Option 9, this option does not offer the least risk to the ports and would result in no Federal dredge capability. Option 7 would result in a $10.1 million net reduction in the total cost for hopper dredging, offer approximately 55 days additional work to industry, ensure that there is a viable reserve capability ready to respond to unforeseen requirements, and ensure that Federal projects requiring hopper dredging can be accomplished in a timely manner and at reasonable cost.

Keeping the *Wheeler* in ready reserve is critical to ensure a capability is available and not scheduled when unforeseen requirements occur. The ready reserve process has successfully worked since 1997, and continuation, is therefore recommended. Scheduling the *Wheeler* for 180 days would offer very little additional capability to respond to unforeseen and peak workload requirements. Putting the *Wheeler* back to work will substantially impact industry workload. The addition of the 12,000 cubic yard hopper dredge, *Glenn Edwards*, scheduled for launching in November 2005, will add an additional 18% capability to the industry fleet, and current excess capability does not justify changing the ready reserve status of the *Wheeler*.

Operating the *Essayons* for 215 days ensures the ability to maintain the West Coast navigation projects with the anticipated longer transit distances to ocean disposal sites, and will not impact west coast industry workload requirements.

Increasing the workload of the *Yaquina* beyond the Status Quo will result in this small class hopper dredge performing work normally accomplished by medium class hopper dredges in deep draft navigation projects. This would not be the most cost effective manner to maintain the ports, harbors and channels of the Pacific coast.

Retiring the *McFarland* will result in some lost capability, however the addition of Manson's 12,000 cubic yard hopper dredge, *Glenn Edwards,* scheduled for launch in November 2005, will more than offset this capability. Other types of dredges have historically performed the majority of the work in the Delaware River, and it is expected that sufficient industry hopper dredge capability exists to perform the requirements that may occur in the Delaware River. The *McFarland* generally dredges spot shoals in the Delaware River, and seeks to keep full project dimensions at all times. Even if the scheduled work for the *McFarland* were maximized, the reduction in daily rate would still be almost double the daily rate of a comparable industry hopper dredge. From a business case, the *McFarland* is the oldest dredge in the fleet, and operates at a daily rate that substantially exceeds comparable industry medium class hopper dredges. If the McFarland were to be kept in the Minimum Fleet it would have to be rehabilitated and repowered at a cost of approximately $20 million. If Congress accepts these recommendations, the Corps would propose a timeline of two years for retirement of the McFarland. There will be a substantial impact to the Philadelphia District staff if the *McFarland* is retired. Approximately 80 people will be directly impacted. Every effort would be made to minimize the impacts to the affected employees. One-time retirement costs will require an additional $22 million be budgeted.

Baseline

The baseline for determining the appropriate use of the Corps hopper dredge fleet is based on a risk management approach and balances operational flexibility and response with maintenance of a viable private hopper dredge industry capability. The basic tenet for establishing the baseline is the language in Public Law 95-269, "*...the Secretary may retain so much of the federally owned fleet as he determines necessary, for so long as he determines necessary, to insure the capability of the Federal Government and private industry together to carry out projects for improvements of rivers and harbors.*" Appropriate use of Corps hopper dredges must be predicated on a sound business case, and could include fully operating a Federal hopper dredge, or placing the dredge in a ready reserve status. The goal should be to identify the best business case that can ensure the ability to provide an acceptable level of service to the ports and users of the Nation's waterborne transportation system. As a result of the analyses and consideration of all the options, the baseline should schedule work for two Corps hopper dredges on the West Coast, maintain a Corps hopper dredge in the Gulf in Ready Reserve, continue to maintain the Industry Corps Hopper Dredge Management Group, seek to schedule approximately 450 days of work for Corps hopper dredges, and keep Government costs under approximately $40 million. The Corps must monitor the capacity of the industry fleet, and ensure that work can be accomplished in a timely manner and at reasonable cost, planned workloads do not exceed capability, and some excess industry and Corps capability is maintained for unforeseen requirements and peak workload demands. The reliability and safety of our Nation's ports and harbors must be maintained.

The following table, Table 13, portrays the annual workload for industry and the Corps by quantity dredged and by workdays. The graph also tracks the total Corps capacity and total industry capacity from FY 2000 through FY 2004. The total capacity for the Corps hopper dredges is 885 days as reflected in Option 1. The total industry capacity is estimated by adding a reasonable maximum capability as reflected in the actual workloads of the individual industry hopper dredges. In some cases the maximum workload of an industry dredge may be unusually high, and a previous years maximum may be used to reflect a realistic maximum. While this approach is not exact, it develops a reasonable baseline ceiling to manage potential capability versus actual workload. With the launching of the additional 12,000 cubic yard hopper dredge, scheduled for November 2005, there will be even more surplus capability.

Table 13. Comparison of Total Hopper Quantity to Industry Quantity and Corps Total Days Worked

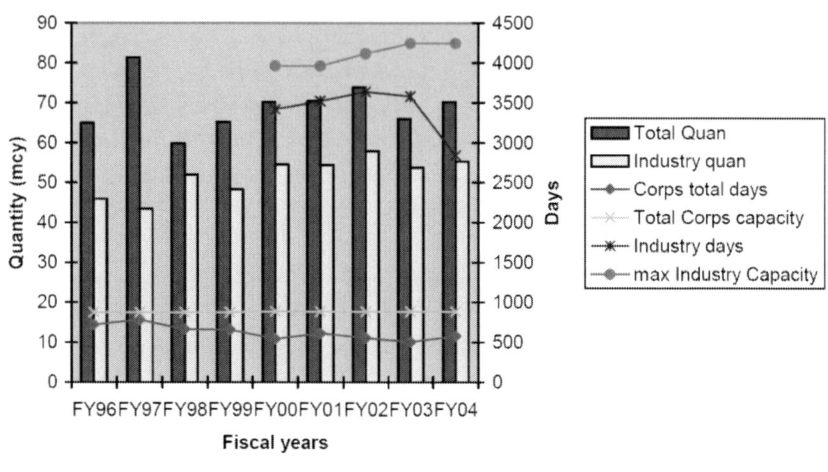

Appendix A. Detailed Options Analysis

Option – Status Quo

Wheeler in ready reserve with 55 training days, McFarland restricted to 85 days, Yaquina works 178 days, and Essayons works 185 days				
Dredge	**Days**	**Daily Rate**	**Subsidy**	**One Time Costs**
Wheeler	55	$75,000	$8 Million	0
Mcfarland	85	$65,000	$7 Million	$20 Million Rehab
Essayons	185	$87,000	0	0
Yaquina	178	$47,000	0	0
Government Costs & Usage		$49.1 million	503 Days	
Contract Costs		$118.7 million		
Total Program Costs		$167.8 million		
Risk				
Navigation	Low			
Industry	Low			
Comments				

(Note: McFarland and Wheeler daily rates held at previous daily rates prior to reduced operating restrictions).

Option – 1. All Corps dredges work maximum number of days

Dredge	Days	Daily Rate	Subsidy	One Time Costs
Wheeler	245	$68,415	0	0
Mcfarland	230	$64,000	0	0
Essayons	215	$76,607	0	0
Yaquina	195	$44,655	0	0
Government Costs & Usage		$56.7 million	885 Days	
Contract Costs		$100.5 million		
Total Program Costs		$157.2 million		
Risk				
Navigation	Moderate Risk			
Industry	High Risk			

Comments: Uncertain workload may result in lost capability of industry dredges. Ability to respond to peak workload demands would be diminished.

Option – 2. FY 2004 Experience. Wheeler in ready reserve, McFarland worked 140 days, Essayons worked 192 days, Yaquina worked 178 days

Dredge	Days	Daily Rate	Subsidy	One Time Costs
Wheeler	55	$75,000	$8 million	0
Mcfarland	140	$87,925*	0	$20 Million Rehab
Essayons	192	$84,304	0	0
Yaquina	178	$47,000	0	0
Government Costs & Usage		$47.8 million	565 Days	
Contract Costs		$116.8 million		
Total Program Costs		$164.6 million		
Risk				
Navigation	Low			
Industry	Low			

Comments: If this option were chosen, the McFarland would be repowered and rehabbed at a cost of approximately $20 million from revolving fund account (PRIP).

*(Note: Actual daily rate for McFarland in FY 04 was $79,500 due to carryover of surplus funds.)

Option – 3. Wheeler works 180 days, Essayons works 215 days, Yaquina works 178 days, and McFarland retired

Dredge	Days	Daily Rate	Subsidy	One Time Costs
Wheeler	180	$88,150	0	0
Mcfarland	0	$87,925	0	$22 million retirement costs
Essayons	215	$76,607	0	0
Yaquina	178	$47,000	0	0
Government Costs & Usage		$40.7 million	563 Days	
Contract Costs		$112.6 million		
Total Program Costs		$153.3 million		
Risk				
Navigation	Moderate Risk			
Industry	Moderate Risk			

Comments: No subsidy for Corps dredges, full crew required for Wheeler, Wheeler work will impact industry workload, no ready reserve capability to respond to unforeseen peak workload requirements

Option – 4. Operate Wheeler, Essayons and Yaquina maximum days, Retire McFarland

DREDGE	Days	Daily Rate	Subsidy	One Time Costs
Wheeler	245	$68,415	0	0
Mcfarland	0	$87,925	0	$22 million retirement costs
Essayons	215	$76,607	0	0
Yaquina	195	$44,655	0	0
Government Costs & Usage		$41.9 million	655 Days	
Contract Costs		$107.7 million		
Total Program Costs		$149.6 million		
Risk				
Navigation	Moderate Risk			
Industry	Medium Risk			

Comments: Reduced West Coast workload for industry, increased excess industry capability, Wheeler workload substantially impacts industry

Option – 5. Wheeler and Mcfarland Work 140 Days, Essayons Works 185 Days, Yaquina Works 178 Days

DREDGE	Days	Daily Rate	Subsidy	One Time Costs
Wheeler	140	$99,500	0	0
Mcfarland	140	$87,925	0	$20 million rehab
Essayons	185	$87,000	0	0
Yaquina	178	$47,000	0	0
Government Costs & Usage		$50.7 million	643 Days	
Contract Costs		$111.2 million		
Total Program Costs		$161.9 million		
Risk				
Navigation	Moderate Risk			
Industry	Moderate Risk			

Comments: High daily rates for Wheeler and McFarland, reduced crew operations limits performing additional work, industry workload reduced.

Option – 6. Wheeler and Mcfarland in Ready Reserve, Essayons Works 215 Days, Yaquina Works 178 Days

Dredge	Days	Daily Rate	Subsidy	One Time Costs
Wheeler	55	$75,000	$8 million	0
Mcfarland	85	$65,000	$7 million	$20 million rehab
Essayons	215	$76,607	0	0
Yaquina	178	$47,000	0	0
Government Costs & Usage			$49.5 million	533 Days
Contract Costs			$118.7 million	
Total Program Costs			$168.2 Million	
Risk				
Navigation		Low		
Industry		Low		

Comments: $15 million subsidy for ready reserve vessels.

Option – 7. Wheeler in ready reserve, McFarland retired, Essayons works 215 days, Yaquina works 178 days

Dredge	Days	Daily Rate	Subsidy	One Time Costs
Wheeler	55	$75,000	$8 million	0
Mcfarland	0	$87,925	0	$22 million retirement costs
Essayons	215	$76, 607	0	0
Yaquina	178	$47,000	0	0
Government Costs & Usage		$37.0 million	448 Days	
Contract Costs		$120.8 million		
Total Program Costs		$157.7 million		
Risk				
Navigation		Low		
Industry		Low		

Comments: Low Government costs, increased work for industry.

Option – 8. Wheeler, Essayons and Yaquina in ready reserve, McFarland is retired

Dredge	Days	Daily Rate	Subsidy	One Time Costs
Wheeler	55	$75,000	$8 million	0
Mcfarland	0	$87,925	0	$22 million retirement costs
Essayons	55	$87,000	$11.3 million	0
Yaquina	55	$47,000	$5.8 million	0
Government Costs & Usage		$36.6 million	165 Days	
Contract Costs		$132.9 million		
Total Program Costs		$169.5 million		
Risk				
Navigation	Medium Risk			
Industry	Low			

Comments: $25.1 million subsidy for ready reserve vessels, substantial additional work for industry, will require an additional large class industry hopper dredge on West Coast.

Option – 9. Retire all Corps dredges

Dredge	Days	Daily Rate	Subsidy	One Time Costs
Wheeler	0	$75,000	0	$65 million retirement costs
Mcfarland	0	$87,925	0	$22 million retirement costs
Essayons	0	$87,000	0	$69 million retirement costs
Yaquina	0	$47,000	0	$22 million retirement costs
Government Costs & Usage		0	0 days	
Contract Costs		$142.6 million		
Total Program Costs		$142.6 million +		
Risk				
Navigation	High Risk			
Industry	Medium Risk			

Comments: $178 million retirement costs for Corps dredges, major impact to PRIP revolving fund account income, Industry must build excess capability, take all risk, navigation projects vulnerable, does not comply with Minimum Fleet legislation. Contract/Program costs will probably be more.

Option – 10. Wheeler and Yaquina in ready reserve, Essayons works 215, McFarland retired

Dredge	Days	Daily Rate	Subsidy	One Time Costs
Wheeler	55	$75,000	$8 million	0
Mcfarland	0	$87,925	0	$22 million retirement costs
Essayons	215	$87,000	0	0
Yaquina	55	$47,000	$5.8 million	0
Government Costs & Usage		$36.6 million	325 Days	
Contract Costs		$125.2 million		
Total Program Costs		$161.8 million		
Risk				
Navigation	Low			
Industry	Low			

Comments: $13.8 million subsidy for Corps ready reserve vessels, increased workload for industry.

OPTION – 11. Wheeler is in ready reserve, Essayons works 185 days, Yaquina works 178 days and McFarland is retired

Dredge	Days	Daily Rate	Subsidy	One Time Costs
Wheeler	55	$75,000	$8 million	0
Mcfarland	0	$87,925	0	$22 retirement costs
Essayons	185	$87,000	0	0
Yaquina	178	$47,000	0	0
Government Costs & Usage		$36.6 million	418 Days	
Contract Costs		$120.8 million		
Total Program Costs		$157.4 million		
Risk				
Navigation	Low			
Industry	Low			
Comments				

APPENDIX B. COST OF OPTIONS

Option	Vessel	Days	Daily Rate	Annual Cost	Subsidy	Gov't Costs ($M)	Contract Days	Daily Rate	Contract Costs ($M)	Total ($M)
Q	Wheeler	55	$75,000	$4,125,000	$8,000,000	$12.1				
Q	Essayons	185	$87,000	$16,095,000		$16.1				
Q	Yaquina	178	$47,000	$8,366,000		$8.4				
Q	McFarland	85	$65,000	$5,525,000	$7,000,000	$12.5				
					$15,000,000	**$49.1**				**$167.8**
1	Wheeler	245	$68,415	$16,761,675		$16.8	-190	$65,700	-$12.5	
1	Essayons	215	$76,607	$16,470,505		$16.5	0	$50,705	$0.0	
1	Yaquina	195	$44,655	$8,707,725		$8.7	-17	$35,800	-$0.6	
1	McFarland	230	$64,000	$14,720,000		$14.7	-145	$34,905	-$5.1	
						$56.7			**-$18.2**	**$157.2**
2	Wheeler	55	$75,000	$4,125,000	$8,000,000	$12.1	0	$65,700	$0.0	
2	Essayons	192	$84,304	$16,186,368		$16.2	0	$50,705	$0.0	
2	Yaquina	178	$47,000	$8,366,000		$8.4	0	$35,800	$0.0	
2	McFarland	140	$87925	$11,130,000		$11.1	-55	$34,905	-$1.9	
					$8,000,000	**$47.8**			**-$1.9**	**$164.6**
3	Wheeler	180	$88,150	$15,867,000		$15.9	-125	$65,700	-$8.2	
3	Essayons	215	$76,607	$16,470,505		$16.5	0	$50,705	$0.0	
3	Yaquina	178	$47,000	$8,366,000		$8.4	0	$35,800	$0.0	
3	McFarland	0	$87,925	$0		$0.0	60	$34,905	$2.1	
						$40.7			**-$6.1**	**$153.3**

(Continued)

Option	Vessel	Days	Daily Rate	Annual Cost	Subsidy	Gov't Costs ($M)	Contract Days	Daily Rate	Contract Costs ($M)	Total ($M)
4	Wheeler	245	$68,415	$16,761,675		$16.8	-190	$65,700	-$12.5	
4	Essayons	215	$76,607	$16,470,505		$16.5	0	$50,705	$0.0	
4	Yaquina	195	$44,655	$8,707,725		$8.7	-17	$35,800	-$0.6	
4	McFarland	0	$87,925	$0		$0.0	60	$34,905	$2.1	
						$41.9			**-$11.0**	**$149.6**
5	Wheeler	140	$99,500	$13,930,000		$13.9	-85	$65,700	-$5.6	
5	Essayons	185	$87,000	$16,095,000		$16.1	0	$50,705	$0.0	
5	Yaquina	178	$47,000	$8,366,000		$8.4	0	$35,800	$0.0	
5	McFarland	140	$87,925	$12,309,500		$12.3	-55	$34,905	-$1.9	
						$50.7			**-$7.5**	**$161.9**
6	Wheeler	55	$75,000	$4,125,000	$8,000,000	$12.1	0	$65,700	$0.0	
6	Essayons	215	$76,607	$16,470,505		$16.5	0	$50,705	$0.0	
6	Yaquina	178	$47,000	$8,366,000		$8.4	0	$35,800	$0.0	
6	McFarland	85	$65,000	$5,525,000	$7,000,000	$12.5	0	$34,905	$0.0	
					$15,000,000	**$49.5**				**$168.2**
7	Wheeler	55	$75,000	$4,125,000	$8,000,000	$12.1	0	$65,700	$0.0	
7	Essayons	215	$76,607	$16,470,505		$16.5	0	$50,705	$0.0	
7	Yaquina	178	$47,000	$8,366,000		$8.4	0	$35,800	$0.0	
7	McFarland	0	$65,000	$0		$0.0	60	$34,905	$2.1	
					$8,000,000	**$37.0**			**$2.1**	**$157.7**

Option	Vessel	Days	Daily Rate	Annual Cost	Subsidy	Gov't Costs ($M)	Contract Days	Daily Rate	Contract Costs ($M)	Total ($M)
8	Wheeler	55	$75,000	$4,125,000	$8,000,000	$12.1	0	$65,700	$0.0	
8	Essayons	55	$87,000	$4,785,000	$11,300,000	$16.1	130	$59,192	$7.7	
8	Yaquina	55	$47,000	$2,585,000	$5,800,000	$8.4	123	$35,800	$4.4	
8	McFarland	0	$65,000	$0		$0.0	60	$34,905	$2.1	
					$25,100,000	**$36.6**			**$14.2**	**$169.5**
9	Wheeler	0	$75,000	$0		$0.0	55	$65,700	$3.6	
9	Essayons	0	$87,000	$0		$0.0	185	$59,192	$11.0	
9	Yaquina	0	$47,000	$0		$0.0	178	$35,800	$6.4	
9	McFarland	0	$65,000	$0		$0.0	85	$34,905	$3.0	
						$0.0			**$23.9**	**$142.6**
10	Wheeler	55	$75,000	$4,125,000	$8,000,000	$12.1	0	$65,700	$0.0	
10	Essayons	215	$76,607	$16,470,505		$16.5	0	$50,705	$0.0	
10	Yaquina	55	$47,000	$2,585,000	$5,400,000	$8.0	123	$35,800	$4.4	
10	McFarland	0	$65,000	$0		$0.0	60	$34,905	$2.1	
					$13,400,000	**$36.6**			**$6.6**	**$161.8**
11	Wheeler	55	$75,000	$4,125,000	$8,000,000	$12.1	$0	$65,700	$0.0	
11	Essayons	185	$87,000	$16,095,000		$16.1		$50,705		
11	Yaquina	178	$47,000	$8,366,000		$8.4		$35,800		
11	McFarland	0	$65,000				60	$34,905	**$2.1**	
					$8,000,000	**$36.6**			**$2.1**	**$157.4**

APPENDIX C. COMMENTS TO THE DRAFT REPORT

The draft report was sent to all interested parties requesting a review of the facts and submittal of comments. The comments received can be categorized into general topics as follows:

Aging Industry Fleet

There are several industry hopper dredges that are in excess of 20 years old, and several commenters expressed concern about potential loss of capability of these dredges. Most of the older dredges have been substantially rehabilitated, and are expected to be fully functioning dredges for some time. If there is a need to replace, or add additional industry hopper dredges, industry has demonstrated that this can be done.

a. What factors go into making the decision to replace antiquated equipment? There are many factors that are considered when making additional capital investments into a market. The industry looks at:
   i. Existing capacity.
   ii. Age of existing capacity.
   iii. Efficiency of existing equipment and what gains in efficiency might be realized with new capital investments.
   iv. Stability of the marketplace, i.e. funding sources, demand for services, volatility of the demand and amplitude of the pricing curve.
   v. Likelihood of competitor investments in the same industry.
   vi. Strength of competitors in adverse market conditions.
   vii. Alternative market regions as fall back under poor market conditions.
b. How do investors view what future workload is needed in order to invest in hopper equipment?
   i. This is just one of the factors that are considered. Investors in hopper equipment must take a long-term view, at least 25 years or so. As long as there is a need to continue to transport goods via waterborne means, there will remain a demand for maintenance and improvement of our waterways. There are many mitigating

circumstances that affect this demand, e.g. the increasing limitations placed on equipment by environmental windows. The trend has not indicated a relaxation of those restrictions, but a broadening of their application. This will result in more work performed in fewer available days. This will create spikes in demand and result in fewer days in the year to recoup fixed costs.

c. At what threshold would investors make a decision not to invest, and how would investors view a change in the Corps policy?
   i. The threshold varies with each individual investor. The decision may depend upon how much the investment represents as a percentage of the overall entity. In the U.S., hopper dredges represent a relatively large percentage of the overall balance sheet and at least with one company, could be the balance sheet. Since the Corps is the largest single customer for the industry fleet, the industry lives and dies by the policy decisions of the Corps.
d. How would investors view a change in Corps policy to increase use of the Corps hopper dredges?
   i. There are many decisions made within the Corps that have immediate and direct implication on the industry fleet. One of these decisions is the subject of this report. Since the late 1970s, it has been the intent of Congress through multiple legislative initiatives to have more of the dredging services performed by private industry as industry demonstrated the capability to perform the work in a safe, cost-effective, and timely manner. For over two decades, we, as a national resource, have made incremental strides in that direction and demonstrated overall success every step of the way. Investment in recent years is a direct result of the confidence industry has developed in the Corps to continue this trend as we continued to demonstrate success. To reverse this trend in the face of the successes we have all enjoyed would stymie further investment in the hopper fleet as a minimum and destroy confidence in the market for future investments, especially in light of such demonstrated success. A shift away from the orderly reduction of work by the Corps dredges would be a reversal of a 27-year program of privatization and would send a clear message to potential investors to avoid investment in private sector dredges.

Table C-1. Hopper Dredge Age, Expected Life, and Major Rehabilitation

Dredge/ Private (P) Or Corps (C)	Year Built	Age	Years Remaining Based on 35-Year Expected Life/Year	Approximate Major Refurbishment Date
Atchafalaya (P)	1980	25	10 Years/2015	2000
Bayport (P)	1999	6	29 Years/2034	2019
Columbia (P)	1986	19	16 Years/2021	2006
Dodge Island (P)	1980	25	10 Years/2015	2000
Eagle 1 (P)	1981	24	11 Years/2016	2001
Essayons (C)	1983	22	28 Years/2033 (50-Yr Life)	2008
Glenn Edwards (P)	2005	0	35 Years/2040	2025
Liberty Island (P)	2002	3	32 Years/2037	2022
Lindholm (P)	1985	20	15 Years/2020	2005
Manhattan Island (P)	1977	28	7 Years/2012	1997
Newport (P)	1983	22	13 Years/2018	2003
Mcfarland (C)	1967	38	12 Years/2017 (50-Yr Life)	2005 (Overdue)
Northerly Island (P)	1983	22	13 Years/2018	2003
Padre Island (P)	1981	24	11 Years/2016	2001
R.N. Weeks (P)	1987	18	17 Years/2022	2007
Sugar Island (P)	1979	26	9 Years/2014	1999
Stuyvesant (P)	1982	23	12 Years/2017	2002
Westport (P)	1978	27	8 Years/2013	1998
Wheeler (C)	1982	23	27 Years/2032 (50-Yr Life)	2007
Yaquina (C)	1981	24	26 Years/2031 (50-Yr Life)	2010

Table includes both Corps and private hopper dredges. Corps to fill-in the expected life and major rehabilitation dates.

Industry Investment in Physical Plant

- Industry has made considerable investments in hopper dredging equipment in order to increase capacity to meet demand. Since 1996, the five existing hopper dredge companies have invested or are investing a total of over $ 165 M. in physical plant. This includes building of the *Bayport* in 1999, the *Liberty Island* in 2002, the *Glenn Edwards*, scheduled to be operational by late 2005; productivity enhancements to the Eagle and Stuyvesant; rebuilding of the *Columbia* and *Atchafalaya*; and improvements to the Island Class dredges.

Industry has indicated that the typical major refurbishment occurs at about a vessel's twentieth year and costs about one-third of the vessel's market value. Based on this assumption, 10 of 15 industry vessels have undergone major refurbishment at an estimated cost of roughly $60 million to $75 million.

Industry Data Not Verified

The industry data used in this report is not used to make conclusions, but is used for relative comparisons in the case of daily rates, and for analyses of workload and cost impacts. To verify the industry data would require extensive auditing and is beyond the scope or need of this report.

All Industry Hopper Work being Taken into Account

All industry hopper dredge work was considered in this report, including private work, beach renourishment, new work and channel maintenance dredging.

Daily Rate Comparison

The daily rates for the Corps hopper dredges has mobilization and demobilization included in the rates, while the industry rates generally reflect just the daily operation of the dredge exclusive of mobilization and demobilization. The Dredging Information System (DIS) was queried for these additional costs on industry contracts and it was found that there was a broad range in the costs for mobilization and demobilization. Variables include proximity of the bidders to the work, risk factors, total days on the job, and additional setup and monitoring requirements. These costs ranged from zero to over $1 million. The average for 52 jobs reviewed, exclusive of West Coast contracts, new construction and beach nourishment, was $3,697 per day. The average mobilization for the West Coast was $15,765, and was derived from 7 contracts. This is a qualitative means of developing comparable industry daily costs, but in relative terms, this average could be applied to each of the daily rates reflected in Table 11 above.

Dredging Requirements

Future dredging requirements was an issue raised by several commenters. Dredging requirements in previous years have reflected a mix of hopper dredging to maintain existing navigation inlets and channels, beach renourishment, private dredging, and new work construction and port deepening. The dredging requirements in any given year are dependent upon several variables, but is primarily influenced by budgetary constraints, weather related shoaling both from coastal storms and inland rains and flooding, and environmental windows and restrictions. There will be peak workload requirements that will engage all available hopper dredge capability, and may warrant imposing management decisions that relocate hopper dredges from one project to another. However, increased industry capability has effectively reduced the potential for disruption or lack of hopper dredge availability. There is a potential for increased dredging as a result of some of the deepening work, and this additional requirement will be closely monitored. Increased demand will warrant increased investment by the dredging industry in additional hopper dredge capability. Current budget horizons will substantially limit increased requirements, however, some adjustment of workload to hopper work may be expected.

Establishing a Baseline

A Baseline discussion has been included in the document

Risk

Risks that were considered in developing the options include risk to the projects, ports, and harbors, and risk to the dredging industry. The risks to the projects and ports is considered high if there is not enough capability, or there is a likelihood that dredges may be out of the country or in a lay-up status that requires several weeks to return to service. Risk to the dredging industry is high when there is a likelihood of not having enough work to ensure a viable income to pay fixed costs and properly maintain the dredges. A consistent workload is a key factor when industry seeks investors for constructing new dredges.

Options

Initially, only selected Corps options were presented in the draft report, now all options recommended by commenters have been included.

National Security Issues

There was concern expressed that reduction in current hopper dredge capability could result in the Corps being unable to support Strategic Ports. Monitoring channel conditions and dredging requirements of Strategic Ports has been an integral part of Corps operations and procedures developed in the Industry Corps Hopper Dredge Management Group (ICHDMG) are in place to rapidly respond to any unforeseen requirements that may occur. Consideration of capability requirements for these Strategic Ports has been included in the risk assessment for the ports, harbors and navigation projects.

INDEX

O

P

R

S

T

U

V

W

Y